FAST FURY

Publications International, Ltd.

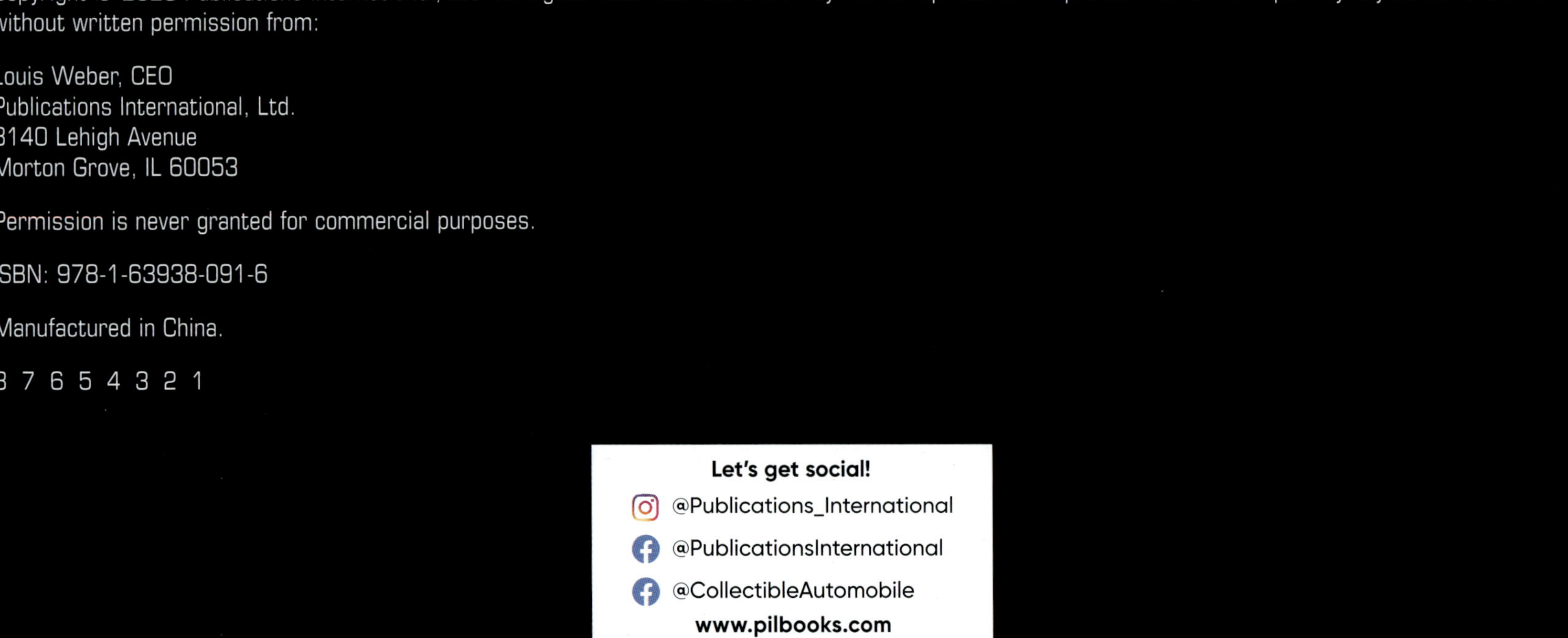

Louis Weber, CEO
Publications International, Ltd.
8140 Lehigh Avenue
Morton Grove, IL 60053

ISBN: 978-1-63938-091-6

Manufactured in China.

8 7 6 5 4 3 2 1

CREDITS

We would like to thank the following vehicle owners and photographers for supplying the images in this book.

2009 Pontiac: O: General Motors Company; P: Jeff Cohn. **2010 Roush:** O: Roush Performance; P: Jeff Cohn.

Special thanks to the following manufacturers who supplied us with imagery.

Aston Martin Lagonda Limited, Audi AG, BMW Group, Bugatti Automobiles S.A.S., Daihatsu Motor Co., Ltd., Daimler AG, Ferrari S.p.A., Ford Motor Company, General Motors Company, Group Lotus, plc., Honda Motor Co., Ltd., Jaguar Land Rover Limited, Koenigsegg Automotive AB, Automobili Lamborghini S.p.A., McLaren Automotive, Mitsubishi Motors Corporation, Morgan Motor Company, Nissan Motor Corporation, Panoz, LLC, Polestar, Porsche AG, Rimac Automobili, Roush Performance, Saleen Automotive, Spyker N.V., Stellantis, Tesla, Inc., Toyota Motor Corporation

Additional images from Shutterstock.com

TABLE OF CONTENTS

PANOZ
ESPERANTE 2000

The product of one of America's smallest viable auto manufacturers, the Panoz Esperante boasted a Ford powertrain and an all-aluminum chassis. A version of the same engine found in some Mustangs, the 4.6-liter V8 produced 320 bhp. Esperantes were expensive and rare. Introduced in 2000 with an $80,000 price tag, fewer than 200 were produced annually. Under hood stainless-steel plates bore the signatures of the craftsmen who hand assembled each car.

panoz
Esperante

JAGUAR

XKR SILVERSTONE 2001

In 2001, Jaguar brought out a Silverstone edition of its grand touring XKR coupe and convertible. To XKR's equipment the Silverstone added: Platinum Silver paint, "Silverstone" writ on hood emblem and chrome door-sill plates, bird's-eye maple interior planking instead of the usual burled walnut, charcoal leather upholstery with red stitching, larger-diameter all-disc brakes by Italy's Brembo, and 20-inch wheels wearing bigger tires. The coupe chassis was further upgraded with a Performance Handling Pack comprising a slightly larger front antiroll bar, a slightly smaller rear bar, higher-rate springs, and steering with a recalibrated electronic control unit and a rack mounted on firmer bushings.

Named for the famed British airfield circuit where Jaguars raced and won, the Silverstone was an evolution of the XK-Series, which bowed as the 1997 XK8 with Jaguar's then-new 32-valve, 4.0-liter, twincam V8 and styling with strong overtones of Jaguar's storied E-Type sports cars. The front suspension and part of the floorpan also looked to the past, being held over from the superseded XJS that dated back to 1975.

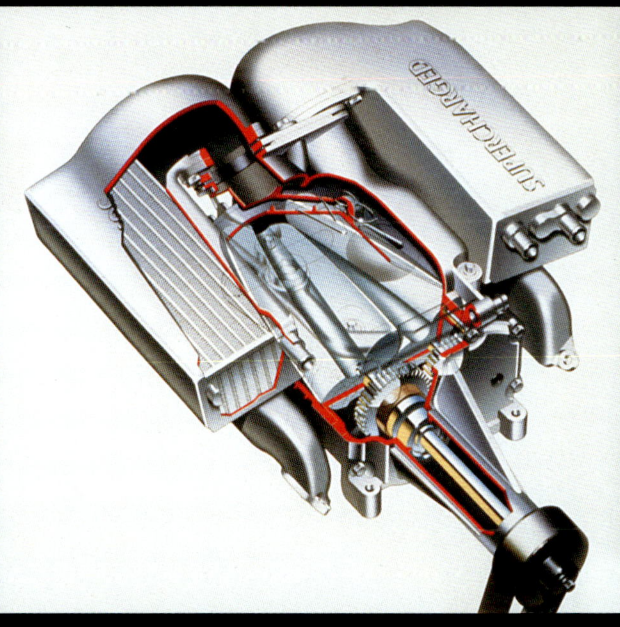

The XK8 was too heavy and luxurious to be the "sports car" Coventry said it was. But the styling implied otherwise, prompting *AutoWeek's* Pete Lyons to ask, "Is the XK8 really a neo-E? Sorry, no. . . . Then is it just a rebodied XJS? No, it's better, much better. The lively new XK8 can honestly call itself a driver's car—a fine modern GT."

Traces of XJS were in it, but the styling was drop-dead gorgeous; handling was enjoyably agile and the V8 was a silky, whispering dynamo. But some people, including most journalists, thought the XK8 deserved more than 290 bhp. Coventry added a blown V-8 for 2000 and created the XKR, that packed 370 bhp and 387 pound-feet of torque. The supercharger made a muted, but discernible, full-throttle moan that recalled the bellow of classic prewar supercharged machines.

Despite its sporty elegance and ineffable "Olde English" Jaguar charm, the XKR proved an exhilarating ride. Most published road tests listed 0–60 mph at about 5.3 seconds regardless of body style. The standing quarter mile came up in as little as 13.7 seconds at 105 mph. Top speed was electronically limited to 156 mph, but disconnecting the governor might have increased that to 170. *Car and Driver's* test coupe generated 0.88g on the skidpad, not Corvette grippy perhaps, but impressive for a fast, smooth-riding GT weighing nearly two tons.

All these stats naturally applied to Silverstones, which weighed no more than like-equipped XKRs. Even so, the bespoke models had an edge in braking (somewhat shorter) and handling (even more adjustable). After testing one for its December 2000 issue, *Road & Track* reported "the Silverstone feels more surefooted [on the track] than the stock XKR, exhibiting less side-to-side rolling when pushed hard around tight corners. A simple lifting of the throttle will tuck the front end right back on track to the apex. And staying on the throttle a bit longer out of a corner will easily invoke a light but controllable progression of understeer to help position the car."

LAMBORGHINI

MURCIÉLAGO 2002

Next in line to carry the Lamborghini supercar torch, Murciélago took over where the Diablo left off. New for 2002, a roadster joined the line in 2004. Though Murciélago means "bat" in Spanish, the stealthy implications of the name were lost on these cars. The beefy Lambo V-12 returned for Murciélago duty, enlarged to 6.2-liters, up from the Diablo's 6.0. Producing a prodigious 575 bhp, the Italian sports car reached 60 mph in a scant 3.7 seconds, according to *Motor Trend*. Top speed was in excess of 200 mph. Rare among exotics, Murciélago put power to the ground through a full-time all-wheel drive system. Murciélago prices started at $280,000.

CF 789KA
BO

AUDI RS 6 2003

The 2003 Audi RS 6 was a one-and-done treat for North America—and even then, squeezed out in a cluster of not quite 1,000 cars. The RS 6 was a high-performance version of Audi's midsize A6 sedan that was every bit the match for the titans from Munich, Stuttgart, and other Old World carmaking capitals. During its whirlwind tour of America, the RS 6 mightily impressed automotive writers with lightning acceleration and excellent all-wheel drive grip—the latter a feature that set it apart from its nearest competitors.

Craft-built by Audi's performance-car subsidiary, quattro GmbH, the 6 sedan was the first RS model to make it to America. (There have subsequently been others.) The thumping heart of Audi's hot rod sedan was an enhanced version of the company's 4.2-liter aluminum V-8. With twin overhead camshafts and five valves per cylinder, it made up to 340 bhp in lesser naturally aspirated models. But with twin intercooled turbochargers and sodium-cooled exhaust valves installed, the RS 6 version delivered 450 bhp at 5,700 rpm. Torque amounted to 415 pound-feet that kicked in at a rock-bottom 1,950 rpm and stayed stout all the way to 5,600 revs. Among its closest imported power-sedan peers at the time, only the Mercedes-Benz E55 AMG could top it in these measurements.

The RS 6 possessed a powertrain that wowed critics by propelling—and that truly is the right verb—it to 0–60 mph runs of as low as 4.3 seconds, making it a near match for the E55, which had nineteen more horsepower and 101 more pound-feet of torque. Quarter mile times were in the high twelve-second zone. North America-bound models had a top speed limited at 155 mph. "The RS 6's disgorging of all its 450 horsepower is so effortless, so jetlike, that you quickly find yourself hurled into a realm where no one can hear you scream," *Automobile* said following an early test.

The suspension got a hand in doing its work from a self-adjusting system that Audi called Dynamic Ride Control. Hydraulic fluid from the shock absorbers was distributed from shock to shock. This rebalancing of fluid was designed to counteract any tendency to pitch in cornering or squat and dive in braking and accelerating. Grip from the standard quattro all-wheel drive and the 255/40ZR18 Pirelli PZero Rosso tires garnered kudos. Skidpad figures neared .9g and bested all comers in various comparison tests. However, a couple of reports mentioned palpable understeer.

on performance couldn't take away from the fact that the RS 6 was a luxury car, too. Standard Nappa leather sports seats were heated, as were the rear seats. The blend of qualities found within the RS 6 made it a critical hit. In a September 2003 head-to-head comparison against the E55 AMG, *Motor Trend* deemed the Audi the better car overall. Four months earlier, *Car and Driver* had ranked it the best in a field that also included the E55 AMG, the BMW M5, and the supercharged Jaguar S-type R. "[T]he RS 6's blend of power, high-speed aplomb, comfort, and superior workmanship carries the day," *C/D* said.

DODGE VIPER 2003

Redone for 2003, Dodge's brutal Viper got a new body and an additional helping of power. Viper's 8.0-liter V10 grew to 8.3 liters, adding 50 bhp for an even 500. Performance was startling, with a 0-60 time of 4.1 seconds and a top speed approaching 180 mph. A single SRT-10 convertible model replaced the previous generation's RT/10 droptop and GTS coupe. Refinements for 2003 included a 100-pound weight reduction, a 2.3-inch longer wheelbase, and a return to the signature side exhaust outlets that disappeared in 1996. A six-speed manual was the standard and only transmission available. Base price in 2003: $79,995

VIPER

FERRARI ENZO 2003

The crown jewel of the Ferrari lineup, the limited-production Enzo came and went in the blink of an eye. Applying the company's famous "demand minus one" formula, Ferrari built just 399 of the stunning coupes—all in 2003 and 2004. Extracting 660 horsepower from a 6.0-liter midship-mounted V12, and weighing less than 3000 pounds, performance was eye-popping. According to *Road & Track* magazine, the Enzo sprang from 0-60 mph in a scant 3.3 seconds, and topped out at nearly 220 mph. Equally breathtaking was the Enzo's price, about $650,000.

The only available transmission was a 6-speed clutchless "sequential" manual unit with steering-wheel-mounted paddle shifters. Extensive use of exotic materials helped make Ferrari's lithe dancer the welterweight it was. Body panels were formed of a carbon fiber and Nomex "sandwich," while the chassis and tub were formed of carbon fiber. Named for the company's founder who died in August 1988, the car's official name was Ferrari Enzo Ferrari, but fans and the press quickly reduced it to simply Enzo. Color choices were limited to yellow, black, and Ferrari's trademark red. Guaranteeing the Enzo's rarity stateside, Ferrari shipped only 100 cars to America.

SALEEN

S7 2003

Best known for its aggressively modified Mustangs, Saleen Engineering entered the production sports car business in 2003 with the wildly extroverted S7. A true American supercar, the S7 boasted a massive 7.0-liter, Ford-based V8, and a trim 2800-pound curb weight. Performance was predictably stunning—*Car and Driver* drove an S7 to 60 mph in 3.3 seconds. Price: $395,000.

MITSUBISHI

LANCER EVOLUTION 2003-2004

The Mitsubishi Lancer Evolution had a rabid international fan base and a storied competition history long before most American enthusiasts had ever heard of it. The Evo wouldn't arrive on our shores until 2003. Back in 1990, Mitsubishi was hungry for a World Rally Championship victory and decided that its best hopes for WRC success would lie in the compact chassis of the Lancer sedan. Engineers set about shoehorning an all-wheel drivetrain into the small Lancer platform to create the first Evolution. To satisfy homologation rules, Mitsubishi was required to offer street versions to the public, so the all-wheel drive, 247 bhp Lancer Evolution I debuted in Europe and Japan in 1992. All through the nineties, ever-more sophisticated Evolution rally cars racked up an impressive string of victories in World Rally Championship and Asia Pacific Rally Championship competition.

The 2003 Evolution was your basic, bread-and-butter Lancer sedan, only it was stuffed to the gills with high-tech performance hardware. Under the hood was Mitsu's veteran 4G63 inline four-cylinder. In the Evolution, this dohc 2.0-liter was good for 271 bhp at 6,500 rpm and 273 pound-feet of torque at 3,500 rpm with the help of an intercooled twin-scroll turbo that put out a maximum 19 psi of boost.

The boxy styling of the standard Lancer never set anyone's heart afire, but the Evo's add-ons gave it a businesslike performance look and an unmistakable presence. Bulging fenders help covered 17-inch Enkei wheels on sticky Yokohama Advan P235/45ZR17 tires. Beefy Brembo brakes, complete with four-piston calipers up front, supplied tenacious stopping power. The turbo's air-to-air intercooler was plainly visible through the thin mesh of the scowling front fascia. The hood's large vent opening expelled engine heat. An aggressive rear spoiler was perched on the decklid. Mitsubishi also offered a bare-bones Evolution RS that deleted nonessential racing equipment, along with the rear spoiler.

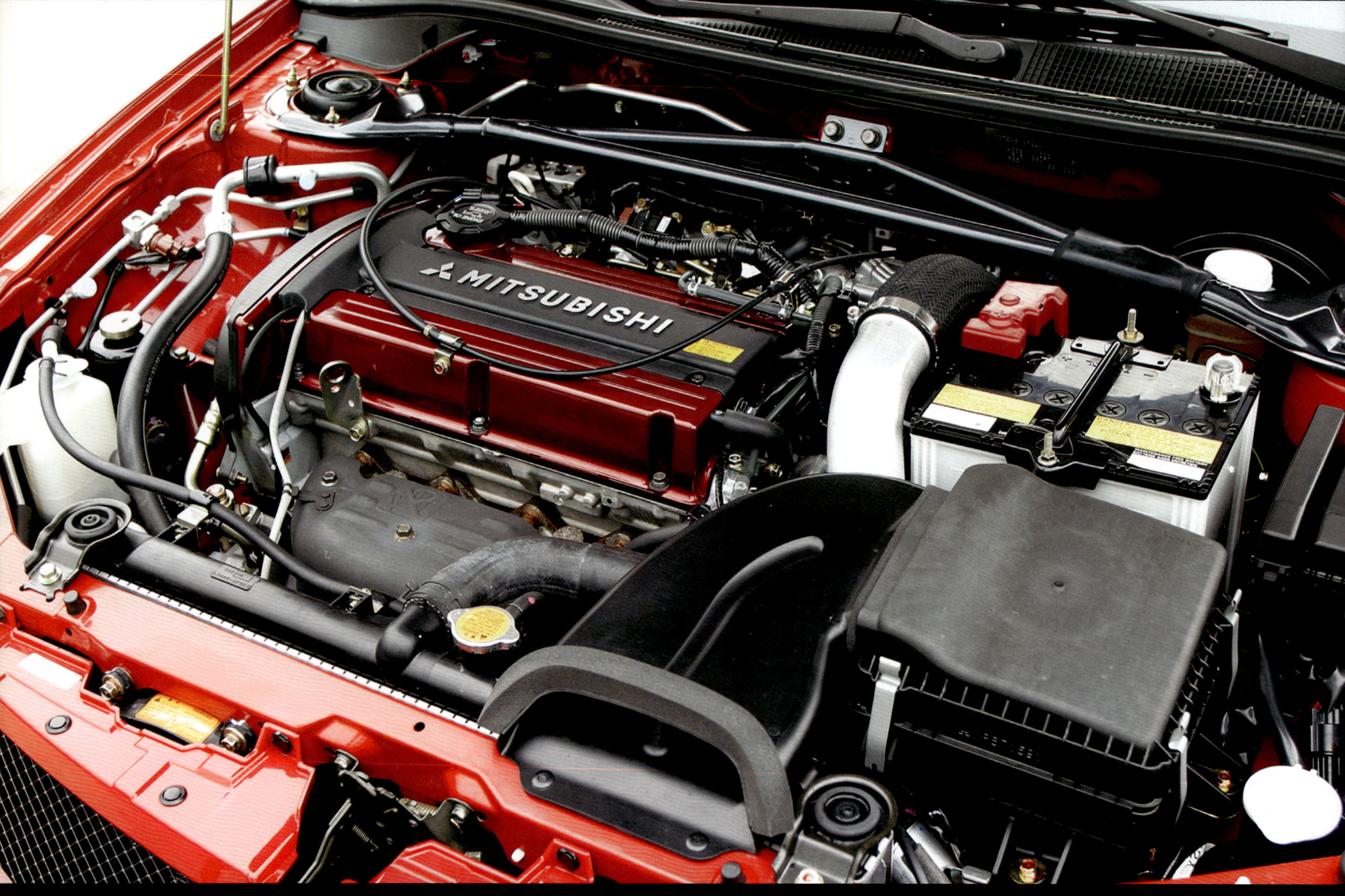

The Evo was about no-holds-barred performance over a variety of road surfaces and in many weather conditions. Acceleration was leisurely up until about 3,500 rpm, when the turbo boost kicked in and the car rushed forward in an intoxicating, hair-raising surge. Mitsubishi's official performance figures were 0–60 in five seconds, the quarter mile in 13.9 seconds, and an estimated 155 mph top speed. These numbers were more than enough to hang with the best muscle cars of yore, and wallop any other sub-$30,000 car of its day. Plus, the Evo's go-kart-quick steering and confidence-inspiring handling made it a joy to drive quickly. The tradeoffs for the extreme cornering prowess were a stiff, occasionally punishing ride that never let you forget that you were in a tautly suspended performance car, and copious amounts of road noise from the low-profile tires.

Auto-enthusiast publications were unanimous in their praise of the Evo's sure-footed moves: "The single most impressive feature of the car is how idiotproof it is, how easy it is to drive really fast," said *Automobile. Automobile* later crowned the Evo its 2004 "Automobile of the Year," going so far as to call it "very likely the automotive icon of the decade."

PORSCHE

CARRERA GT 2004

Bearing a name synonymous with sports cars, Porsche was not about to be left out of the supercar renaissance. The stunning Carrera GT arrived on the scene for 2004, and moved promptly to the head of the Porsche class. The GT boasted Porsche's largest-ever street-going engine, a 5.7-liter V10. With 605 bhp on tap, the midship-mounted engine moved the 3000-pound GT to 60 mph in a factory-claimed 3.9 seconds. Top speed was reported to be in excess of 200 mph. To make the most of the car's limited storage space, Porsche included a matching five-piece luggage set with each car. Price: just under $400,000.

Carrera GT

PORSCHE
PORSCHE

ROUSH
MUSTANG 440A 2004

Roush Performance, of Livonia, Michigan, was one of a handful of aftermarket firms that blossomed from bolt-on-parts supplier into car producer. Their turnkey stallions were assembled using parts from Roush's own high-performance catalog. Roush Performance had already cooked up its own hot version of the new-for-2005 Mustang, but company founder and engineering guru Jack Roush figured that the outgoing 2004 model deserved a special send-off. Hence, the limited-edition Roush 440A.

The 440A nameplate decoded like this: 400 horsepower, the 40th production year of the Mustang, and "A" for anniversary edition. The 40 also represented the production figure for the car—a mere 40 were built. Underhood, the 440A packed Ford's modular 4.6-liter V8. This engine made 260 bhp in a stock 2004 Mustang GT, but in the 440A it cranked out 400 horsepower with the help of a Roush intercooled Roots-style supercharger.

Though the 440A retained a live rear axle and stock Mustang brakes, the chassis received a few tweaks. Roush's Stage 3 suspension setup included specially valved shocks and struts, with Roush lowering springs, dampers, aluminum rear control arms, and a 30mm front antiroll bar. Special Roush chrome five-spoke wheels mounted beefy B.F. Goodrich rubber.

Muscle Mustangs & Fast Fords magazine recorded a best ET of 13.36 seconds at 106.88 mph in a prototype 440A convertible and found the car tight and confidence-inspiring on a few road-course hot laps.

Externally, the 440A got its muscled-up look from a body kit with a unique hood, an aggressive front fascia with integrated fog lamps, rocker panel side skirts, and a rear wing. Shelbyesque racing stripes covered the hood, roof, rocker panels, and decklid.

Sticker prices ranged from $51,000 to $58,000, depending on equipment. In both physical modifications and overall spirit, Roush's Mustangs were considered successors to Carroll Shelby's legendary GT-350s and GT-500s. Jack Roush might not have been quite the household name among enthusiasts that Carroll Shelby was, but his career was no less impressive. Roush had 40 years of hands-on experience with Ford products, and an amazing record of wins in a wide range of competition, from drag racing, to road racing, to NASCAR.

440A

CADILLAC

XLR 2004-2005

The Cadillac XLR that debuted in 2004 was a suave Mercedes SL fighter based on Chevrolet Corvette bones. Yet the XLR looked nothing like a Corvette or the SL, had a pure Cadillac heart, and was plenty fast. The XLR was basically the "for sale" version of the 1999 Evoq concept car, the first public hint that Cadillac's future in the 21st century would not be like Cadillac's past. At the time the Evoq broke cover, General Motors's luxury brand was some two years into a $4 billion extreme makeover dubbed "Art & Science."

The XLR was not a Corvette in a Cadillac suit, though the similarities were undeniable. For starters, both had front-mounted V8s, rear transaxles, and composite-paneled bodies. Apart from looks and intended audience, it was the powertrain that most separated the GM twosome. The C6 Corvette, of course, boasted Chevy's 6.0-liter LS2 ohv V8 with a stout 400 bhp. The XLR predictably used Cadillac's own dual-cam 4.6-liter Northstar V8, only it was a reengineered "Gen II" version developed for rear- and all-wheel drive models vs. the original Northstar's strictly front-drive applications. Highlights included continuously variable cam timing for both intake and exhaust valves, new cylinder heads with free-flow ports and higher compression (10.5:1), stiffer block and crank, low-friction polymer-coated pistons, manifolds redesigned for quieter running, and "by-wire" electronic throttle control. The result of all this was 320 bhp at a zingy 6,400 rpm, and 310 pound-feet of torque at 4,400.

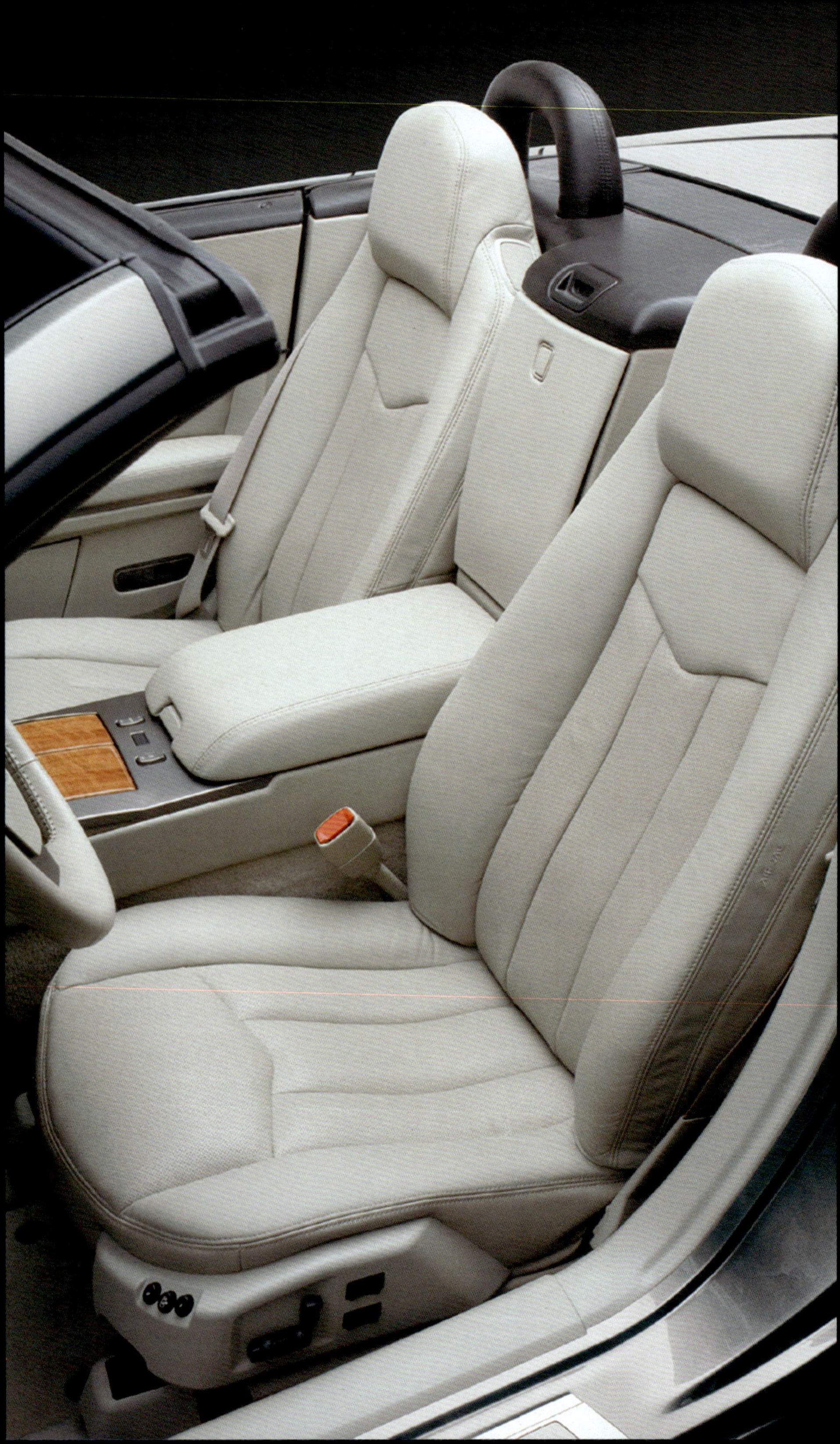

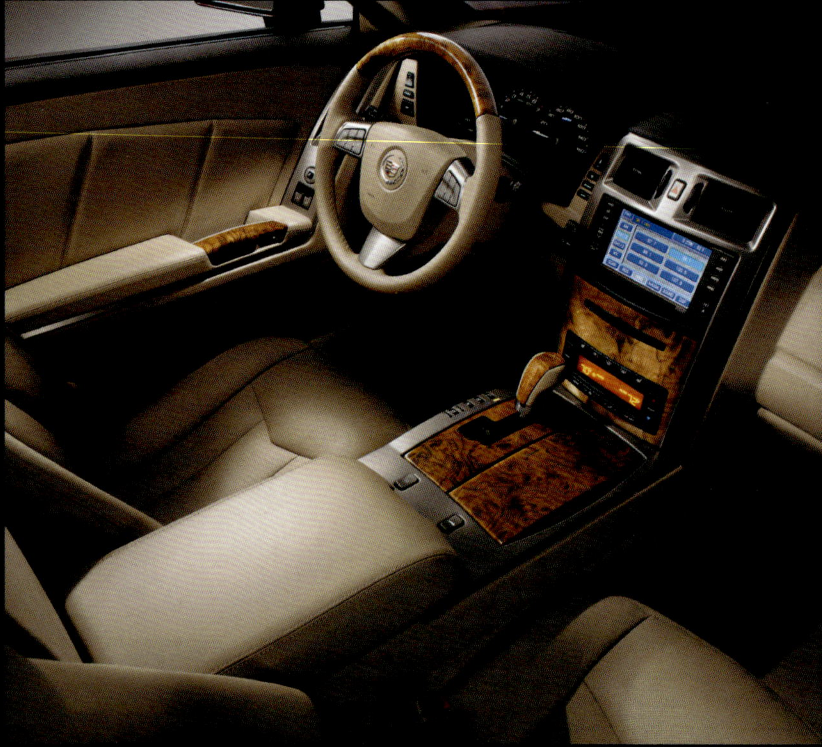

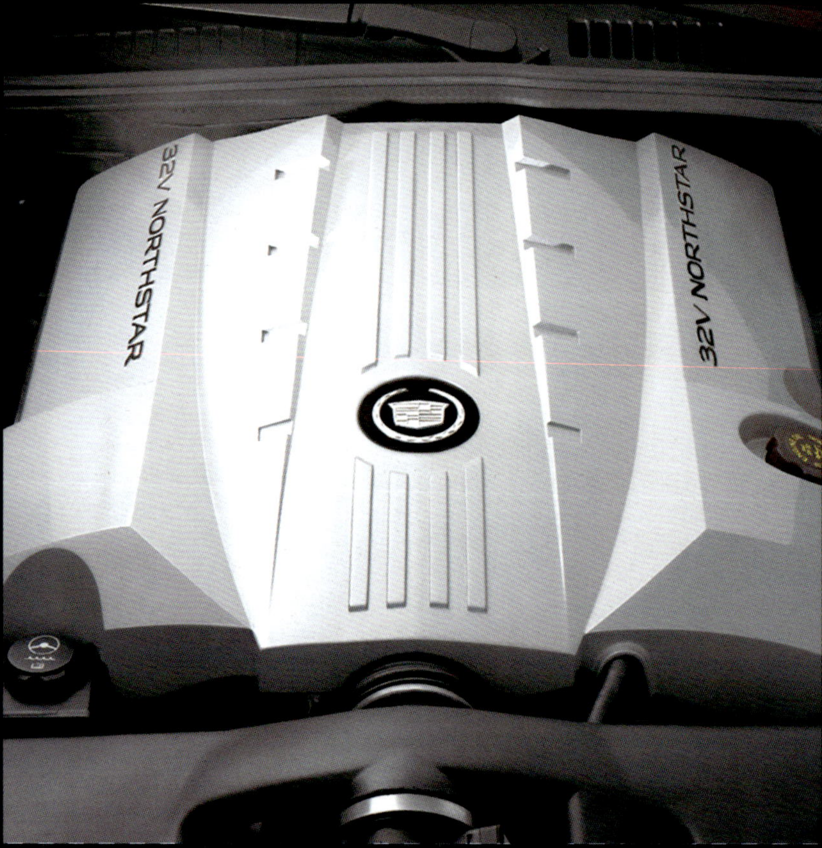
32V NORTHSTAR
32V NORTHSTAR

Another distinction was the XLR's mandatory five-speed automatic transmission with manual shift gate vs. Corvette's four-speed automatic or six-speed manual. The XLR also replaced the 'Vette's conventional convertible top with a disappearing hardtop and added luxury-class equipment.

Most every road test at the time judged the XLR one "Xcellent Luxury Roadster." Indeed, the critics praised most everything about it, from the smooth, quiet acceleration—0–60 mph took no more than the factory-claimed 5.9 seconds—to satisfying backroad agility, controlled but comfortable ride, and terrific build quality. Cadillac worked extra hard on that last one, setting up a separate XLR assembly operation that shared only the Bowling Green paint shop and chassis-welding area with Corvette.

There were two bones of contention, though. The power steering still left something to be desired for precision and feedback, and some deemed the run-flat Michelins a bit narrow and soft-sidewalled for best handling. Then again, the XLR hardly wanted for skidpad grip—*Car and Driver* reported 0.83g, *Road & Track* a fine 0.88. A Mercedes SL500 had more cornering power, but was slower in a straight line and cost some $11,000 more.

In all, the XLR was arguably the most impressive new Cadillac since the '67 Eldorado. "[T]here is no question that [it's] a strong entry in the prestigious roadster class," concluded *C/D*'s Csaba Csere. "Cadillac's march toward luxury credibility has reached another important milestone."

PONTIAC
GTO 2004-2006

The 2004-06 Pontiac GTO shared nothing with the hallowed Goats of yore aside from rear-wheel drive and a thumping V8. That's because it was an Americanized version of the four-seat Holden Monaro coupe from GM's Australian branch. Bob Lutz, celebrated GM vice-chairman and "product czar," looked hard at the Monaro in early 2002 as a sales and image booster for the struggling Pontiac brand—a timely idea with the Firebird ponycar then on the way out. Lutz liked what he saw and soon confirmed the Monaro would come stateside for 2004 as the first GTO in 30 years.

Though it seemed like just a Monaro with a twin-port Pontiac grille and left-hand drive, *Road & Track* noted 475 unique parts, "about 20 percent of the car."

The GTO was rather large for a modern midsize. Curb weight was a burly 3770 pounds, but that was no strain for a 5.7-liter/350-cid LS1 V8 lifted more or less intact from the base C5 Corvette. With 350 bhp and 365 pound-feet of torque on 10.1:1 compression, the Yank outmuscled its Aussie cousin by about 50 horses and 25 pound-feet. Mr. Lutz didn't fool around. A four-speed automatic transmission was standard; $695 bought a sturdy Tremec T56 six-speed manual.

Acceleration was predictably vivid. *Road & Track*'s manual-equipped '04 clocked 0-60 mph in 5.3 seconds, 0-100 in 12.9, and a standing quarter-mile of 13.8 seconds at 103.8 mph, stats to shut down the fastest showroom Goats of the golden age. Yet this was no cart-sprung, limp-wristed rocket that fell to pieces on twisty roads. The new GTO provided assured Euro-style handling with little cornering lean, fine grip—0.81g on the R&T skid-pad—and clear, properly weighted steering, plus braking finesse and mechanical refinement classic GTOs never knew. In fact, this car drove like a posh big BMW coupe playing a sixties-style Motown soundtrack.

Which was precisely the trouble. For aging Goat lovers and many younger critics, this reincarnation, however rapid and tuneful, was just too civilized to be a real GTO. The reborn GTO was quite the high-performance bargain, but sales were sluggish. Sensibly, Pontiac scaled back production for 2005 and added not one hood scoop but two. They didn't connect to a power-boosting Ram-Air setup, but they did help cool a more muscular V8: the new 6.0-liter/364-cid LS2 from the C6 Corvette. Outputs jumped to 400 bhp and 395 pound-feet. A scoopless hood continued as a no-cost option. The 2006 edition got only a few cosmetic tweaks and optional 18-inch wheels. Base price had hardly budged, but demand remained stagnant. With conserving cash now imperative for GM, the GTO was cancelled after the '06 run.

It's a shame the Aussie Goat left so soon. Said *Car and Driver*, "[A]s the song goes, you won't know what you've got till it's gone."

BUGATTI

VEYRON 2005

Bugatti's twenty-first-century renaissance began with Veyron and its promised 1001 horsepower V16. Federalized models actually delivered 987 horsepower, roughly double Viper's power output. Production began for 2005, and was limited to 50 vehicles annually. Now under Volkswagen control, the latest Bugatti revival was free of the financial limitations that killed a 1990s rebirth effort.

BUGATTI

FORD GT 2005

Originally conceived of as a GT40, Ford's 2005 reinvention of the legendary LeMans racer was an all-new supercar simply known as GT. Recalling its 1966 LeMans victory over Ferrari, Ford again set its sights on outperforming the Italian sports car builder, this time aiming to outgun its 360 Maranello. Power targets were impressive, 500 bhp and 500 pound-feet of torque from the GT's supercharged-5.4-liter V8. Intended to be a technological tour de force, the GT's credentials were impressive. Ultrastiff floor panels were formed of a lightweight mesh graphite and aluminum "sandwich." The rigid space frame and body panels were aluminum.

MERCEDES-BENZ

SLR MCLAREN 2005

Recalling the glory of Mercedes' 1950s racing success, the 2005 SLR McLaren probed the limits of front-engine performance. With a body and chassis codeveloped with McLaren Racing Development, and a heavily massaged version of Mercedes' already impressive supercharged V8, the SLR was granted instant supercar status. Weighing in at approximately 3000 pounds, and with more than 600 bhp on tap, performance was breathtaking. The SLR's flip-open doors recalled the "gullwing" arrangement of early SL coupes. High-tech

SLR
S LR 6722

FORD THUNDERBIRD 2005

The original 1955-57 Ford Thunderbird was so sleek, so suave, so seductive. Cue up the movie *American Graffiti*. Director George Lucas knew the magic of those early 'Birds. Why else would he have chosen one for the mysterious blonde who teases one of the lustful teen-age heroes with a fleeting drive-by on cruise night?

The new-millennium 'Bird sought to rekindle that sort of magic for the "happy days" generation, but was also aimed at younger buyers enamored of retro fifties style. Like the original, it was conceived strictly as a posh V8 boulevardier two-seater convertible, not a sports car or muscle machine. If you wanted a hot Ford, you got a Mustang GT or Cobra.

The 2002-05 Thunderbird was built on a modified version of a platform shared with Lincoln LS and Jaguar S-Type—Ford owned Jaguar at the time. It also shared a basic rear-drive V8 powertrain with those sedans. Its 240-cid V8 developed 252 horsepower in 2002 and got a power boost to 280 for 2003-05. *Car and Driver* found useful performance gains, with 0-60 mph falling from 6.9 to 6.5 seconds (sans hardtop) and the quarter-mile run improved from 15.2 seconds at 94 mph to 15 flat at 95. One could argue that Ford should have gone beyond "relaxed sportiness." Memory, after all, is a tricky thing, and though the new T-Bird was always much faster than its classic forebears, it probably didn't seem so in the minds of some over-50s. However, if there was such a thing as "heritage feel," this car had plenty. Comfortable and cute, it was just the ticket for people who liked the idea of a sports car without all that pesky sport, much like the fifties original.

Predictively, the new-century T-Bird boasted amenities never dreamed of in '55: air conditioning with automatic temperature control (dual-zone, yet); keyless-entry power door locks, antitheft system; variable intermittent wipers; and a stereo. Plus, the new 'Bird also had safety features unheard of in the fifties: airbags, traction control, and antiskid system. Styling was of the "heritage design" theme. It had things about it you thought you had seen before without being direct copies of things you know that you had seen before. One recreated styling theme was a lift-off hardtop with portholes. The grille, modest hood scoop, and big, round taillights also harked back to the original T-Bird.

The retro 'Birds' Wixom, Michigan, facility was another tie to Thunderbird's past. It was the same plant that opened back in 1958 to build the first four-seat T-Birds and giant unibody Lincolns. Wixom closed in 2007, by which time it was only building Lincoln Town Cars.

Like most "fashion cars," the retro 'Bird fast fell off the sales perch once early adopters got theirs. Two thousand five was Thunderbird's 50th anniversary and final year. There was a 50th Anniversary commemorative fender chevron on every 2005 Thunderbird, plus a 50th Anniversary Edition trim package was available. Just as the original two-seat T-Bird was short lived, so was the retro 'Bird.

LOTUS ELISE 2005-2011

Lotus cofounder Colin Chapman said, "Adding power makes you faster on the straights, subtracting weight makes you faster everywhere." More than a decade after Chapman's 1982 death, his successors proved they still believed in weight reduction with the flyweight Elise, launched in Europe in 1996. Destined to become the best-selling Lotus ever, the Elise took a few more years to jump the pond to the U.S., but it was worth the wait for drivers who appreciated racer-sharp handling and a high level of car/driver symbiosis.

The mid-engine two-seater tipped the scales at around 1,600 pounds at the time of its debut. When it finally went on sale in the U.S. in mid 2004 as an '05 model, it was as a more crashworthy "Series 2" version that had bulked up, but only in the vicinity of 1,950 pounds

The American Elise was powered by a Toyota 1.8-liter four-cylinder engine from the Celica GT-S that Lotus tuned to produce 190 bhp at 7,800 rpm and 138 pound-feet of torque at 6,800 revs. That output gave it a fairly low power-to-weight ratio of just over 10 pounds per horsepower. The gearbox was the same six-speed manual found in the GT-S.

The enthusiast press in the U.S. was smitten with the federalized Elise from the first. Glowing reviews especially praised its responsive steering, sticky grip, and nearly understeer-free manners. (*Motor Trend* and *Automobile* both invoked images of go-karts.) *Motor Trend* concluded, "The Elise is, by most reasonable measures, the best-handling car you can buy today."

LOTUS
ELISE

The engine got plenty of love, too. "As revs climb and the VVTL-i kicks in, the Elise reacts with an immediate burst of energy accompanied by a more vocal response from the intake and exhaust," noted *Road & Track*. Lotus estimated a 4.9-second 0–60-mph time, but a couple of testers were a few tenths quicker. Quarter mile times were in the low thirteen-second range.

For 2008, a supercharged SC version joined the naturally-aspirated engine. The SC engine was the same modified Toyota four, but with a Magnuson-built Roots-type blower that raised output to 218 bhp at 8,000 rpm and torque to 156 pound-feet at 5,000 rpm—and more than compensated for the fact that the car now weighed 2,006 pounds. Standing-start sprints to 60 mph took about 4.5 seconds, but 13.3 seconds in the quarter mile wasn't any better than the unblown car had gotten a few years earlier.

Styling was tweaked for 2011 and that was the end of the Elise story in the U.S. For one thing, Toyota stopped making the engine. Plus, Lotus's waiver for selling cars in the U.S. without "smart" air bags was destined to expire that summer. Certifying a new engine and doing the engineering necessary to meet the air bag standard was too much for what was essentially a limited-edition car.

While gone from America, Elise was still on sale in markets around the world. An exit from the U.S. market didn't diminish the intrinsic value of the Elise as it was during its years in America. Thrilling driving was thrilling driving—as the car's relative popularity proved.

AUDI
S5 COUPE 2008-2010

Road & Track called the Audi S5 a "hard-charging, road-gripping German tourer of the first order." The S5 went on sale in late 2007 and inaugurated the high-style A5/S5 series that slotted between the compact A4 and midsize A6 groupings. The S5 was the high-performance version of the A5 coupe.

The S5 and A5 were Audi's first sports coupes in 20 years, spiritual successors to the five-cylinder front-drive GT and turbocharged Quattro fastbacks of 1981-91. That first square-rigged Quattro marked the Audi brand's coming of age, winning two drivers' and manufacturers' titles in international rallying and touching off a short-lived craze in AWD road cars that later blossomed anew, particularly among premium brands.

S5 coupes were motivated exclusively by Audi's familiar all-aluminum 4.2-liter (254-cid) twincam V8 with four variably timed valves per cylinder. In that application it delivered 354 bhp right at redline, 7000 rpm, and a healthy 325 pound-feet of torque at 3500, aided by a high 11.0:1 compression ratio made possible by direct fuel injection. Transmission choices involved a six-speed manual or ZF's six-speed Tiptronic torque-converter automatic with manual shift gate and steering-wheel paddles.

The S5's styling was crafted by Italian-born Walter de' Silva, who came to Audi after a stint at Alfa Romeo. It was a handsome coupe: chunky, muscular, and well-proportioned, with adept surface development and well-managed reflections. It would have looked even better without B-posts, but that would have required weighty structural reinforcements and the S5 already scaled a hefty 3795 pounds at the curb. (Quattro got part of the blame, as usual.) One nifty touch was the row of LED running lamps curling beneath each headlamp module like Black Bart's mustache.

The S5 was about the same size as a BMW 3-Series two-door and Mercedes-Benz E-Class coupe, standing 182.5 inches long, 73 inches wide, and 53.9 inches high over a 108.3-inch wheelbase. Like most coupes, it was a cozy 2+2, not a practical five-seater. A full-length center console eliminated the rear center position anyway. This relative lack of total passenger space was somewhat offset by Audi's standard-setting cabin execution. That meant the S5 was as lovely inside as it was outside, with scads of aromatic leather and tasteful dollops of aluminum that could be replaced by optional wood, stainless-steel, or carbon fiber accents.

For all its heft and luxury, the S5 claimed impressive performance stats. Most road tests reported 0-60 mph in a quick 4.8 seconds or so (with manual), standing quarter-mile runs in about 13 seconds at 105 mph, and more than 0.90g of skidpad grip. What these numbers didn't convey was the smooth sophistication with which the S5 got about its business.

The Audi S5 coupe was one of the most refined, capable, and satisfying 2+2s of its time, not to mention one of the prettiest. *Road & Track* rightly called it "a treat for all the senses."

TESLA
ROADSTER 2008-2011

The 2008 Tesla Roadster was the first modern, dedicated, highway-speed, production electric car sold in this country. (GM's EV-1 was only leased). Today, Tesla is selling sedans and SUVs, but the company's first product was a small sports car. The low-production Roadster allowed Tesla to test its cutting-edge engineering before rolling out its higher-volume models.

Tesla is a California-based company cofounded by Elon Musk, a man of big dollars and even bigger dreams who made the bulk of his multibillions through the sale of PayPal in 2002. Early prototypes of the Roadster were displayed in the second half of 2006, with the first production model going to Musk himself in March 2008. It's unclear whether Musk paid the full $109,000 list price (minus a $7,500 tax credit) or got an employee discount.

The Roadster's sleek carbon fiber body was made in France, then shipped to Lotus of England for placement on a specially designed chassis. Although the outcome looked much like a stretched Lotus Elise, Tesla said only about six percent of the parts were shared between the two cars. After assembly, the body/chassis was shipped to Tesla's shop in Menlo Park, California, to have the battery pack (6,831 lithium-ion laptop-computer batteries), powertrain, and electronic controller installed.

The first Roadsters had a 248 horsepower motor. By 2010, horsepower had been increased to 288. Furthermore, a Sport edition was added that had the same 288 bhp but more torque (295 pound-feet versus 273) that dropped the 0–60 time to 3.7 seconds—a figure matched by only a handful of cars, all of which relied on boatloads of gasoline to achieve the desired results. Acceleration was explosive and instantaneous at any speed. There was only one gear, so there was never a wait for the transmission to downshift. Touch the throttle and the Tesla jumped; nail it and it pinned you to your seat. There was little body lean in corners, and the Tesla exhibited seriously sporty moves. Steering was tight and go-kart quick. The Roadster wasn't without faults. The steering, being unassisted, was quite heavy at parking and even around-town speeds. Ingress and egress were cumbersome at best. Once inside, it was a snug fit for two, as adults—even not-particularly wide ones—sat shoulder-to-shoulder. Also, while visibility to the front and sides was fine, there was but a tunnel-vision image in the rearview mirror, and the lengthy rear roof pillars obstructed nearly everything over your shoulders.

For 2011, Tesla introduced the Roadster 2.5. The most noticeable change was a revised front fascia. But the cars also got a new rear diffuser, restyled wheels, more supportive seats, additional sound insulation, and an optional seven-inch touchscreen display with backup camera that helped with backing up.

Perhaps the Tesla Roadster's most profound contribution to automotive history was that it was the first such car to so vividly demonstrate the promise of electric vehicles. The fact that it was staggeringly quick, a thrill to drive, and possibly the ultimate automotive conversation piece was only icing on the cake.

NISSAN GT-R 2009

It might have been the most famous Nissan that you'd never heard of—unless you kept up with international automotive unobtainium, followed the drifting crowd, or exercised your inner Andretti at the virtual controls of your Sony PlayStation. But the GT-R had long been worshipped by the automotive underground even outside its home market of Japan, the only country in which it was officially sold—until 2009.

The GT-R traced its history back to the late sixties. GT-Rs held numerous records at the Pikes Peak Hillclimb and on Germany's Nürburgring. They were also terrors on the street, establishing a performance legacy that lived on—and extended well beyond Japan's borders. There were several gaps in the GT-R lineage with the last ending with the debut of the fifth generation GT-R in the fall of 2008. That GT-R had a 3.8-liter twin-turbo V6 rated at 480 bhp at 6400 rpm and 430 pound-feet of torque starting at 3,200 rpm. Some believed those figures were understated with actual horsepower in 500 to 575 range and torque closer to 490 pound-feet. Transmitting the power was a rear-mounted six-speed "automated manual" dual-clutch transaxle with paddle shifters, and a sophisticated all-wheel drive system.

The Nissan GT-R might well have been the performance deal of its day. Consider these numbers: 3.4, 11.8, 193, and $72,880. Those were the 0–60 mph time, quarter-mile ET, top speed, and as-tested price of a GT-R thrashed by *Road & Track*. Other buff books had posted similar performance figures. It was impossible to have duplicated the first three at anywhere close to the fourth.

One reason for those impressive acceleration numbers was Launch Control. Launch Control worked something like this: Choose the "R" (Race) setting for the transmission and suspension, and switch off the stability control. Select the transmission's manual-shift mode with the gear lever. Hold down the brake pedal while flooring the gas. The engine would rev to about 4,500 rpm, hold there, and when you sidestepped the brake, the car launched "like an arrow from a crossbow," according to *Motor Trend*. The AWD system got at least partial credit, as it momentarily sent some power to the front wheels for better bite off the line. Thereafter, the transmission's quick shifts—which needed to be done manually via the steering-wheel paddles and came up very quickly—resulted in virtually zero power slack until you lifted off the loud pedal.

Although it was a boon to acceleration times, the transmission wasn't exactly a paragon of smoothness. When coming to a stop, the transmission jolted down one gear at a time and occasionally jerked on gentle takeoffs, that made stop-and-go driving a head-wobbling affair. It was a similar story out on the highway. Floor the throttle and you could count as the transmission dropped down one gear at a time to the desired ratio. And a "countdown" it was; when the proper gear was finally summoned, the car exploded into warp drive that had you reaching "Good morning, judge" speeds in very short order.

PONTIAC

G8 GXP 2009

Pontiac went down with flags flying and guns blazing. The Pontiac G8's performance actually exceeded that of cars from Poncho's glorious muscle car years. Built in Australia and based on the Commodore VE by Holden, General Motors's branch "down under," G8 was Pontiac's first large rear-drive sedan since the Parisienne was retired in 1986. But it was also the last widely acclaimed product from Pontiac, a car barely able to get grounded in the market before a financially battered GM pulled the plug on its parent division in 2009.

The G8 wasn't able to save Pontiac, but that shouldn't reduce the car's reputation. It was hailed by enthusiasts and well-received by the automotive press. G8 arrived as a 2008 model in base form with a 256 horsepower 3.6-liter V6 and as a GT with a 361 bhp 6.0-liter V8. The high-performance GXP version arrived for 2009. While most of these Aussie-built Pontiacs were powered by burly V8s, the G8 GXP of 2009 stood out as king of this hill. The G8 GXP packed a slightly detuned Corvette engine under its hood. The 6.2-liter LS3 was rated at 415 bhp and 415 pound-feet of torque—down only 15 ponies from the Corvette original. A six-speed automatic was standard. GXP was the only G8 available with a six-speed manual transmission.

Pontiac cited the GXP's 0–60 mph time at 4.7 seconds, with a quarter mile time of 13 seconds flat at 108 mph. GXP performance came at a price at the gas pump. EPA-estimated fuel economy was thirteen mpg city/twenty highway, numbers that earned GXP a $1,700 gas-guzzler tax.

All G8s used four-wheel independent suspension, but the GXP's was tuned at Germany's famed Nürburgring race circuit. The rubber was 245/40R19 Bridgestone Potenza RE050As mounted on polished-aluminum wheels. *Automobile* magazine said this resulted in one of the most neutral-handling sedans on the market. Stopping power was upgraded from the standard G8 setup as well. Brembo four-piston calipers grabbed 14-inch rotors in front; in the back, single-piston calipers worked on 12.76-inch rotors.

GXP's base price was just over $40,000, including the destination charge and the gas-guzzler tax. This was a nearly $8,000 premium over the already muscular and very capable G8 GT, but the pricing was pretty competitive with Dodge's burly Hemi-powered Charger SRT8.

Following the announcement of Pontiac's imminent closing in late April 2009, G8 production was quickly halted. According to sources at GM, a total of 26,368 2009 Pontiac G8s were produced. Of those, only 1,824 were the top-dog GXP model. Of them, 981 were fitted with the automatic transmission and 843 were built with the six-speed manual. The short-lived Pontiac G8 GXP deserved a spot on lists of the best high-performance Pontiacs of all time. The fact that this very special sedan came at the tail end of the brand's existence—and in very low production numbers—only made the GXP more desirable.

CADILLAC

CTS-V 2009-2012

Cadillac's second-generation CTS-V debuted for '09 packing the most-powerful production engine that Cadillac had ever offered: a revised Chevrolet Corvette ZR1 supercharged and intercooled LS9 6.2-liter V8. The Cadillac variant, dubbed LSA, received unique tuning and components aimed at increasing quietness and refinement, but with little performance sacrifice.

The numbers were jaw-dropping: 556 horsepower at 6,100 rpm, and 551 pound-feet of torque at 3,800 rpm. For comparison, the 2009 Mercedes-Benz E63 AMG was rated at 507 bhp, the '09 BMW M5 at 500 bhp. If the original CTS-V was Cadillac stomping on its velour-interior/vinyl-top/wire-wheel-cover/V8-6-4 past, then the second-gen CTS-V was an emphatic grinding of the boot heel.

CTS-V offered a choice of transmissions: a Tremec TR-6060 six-speed manual or General Motors' Hydramatic 6L90-E six-speed automatic with steering-wheel-mounted buttons for manual shifting.

A handful of performance-themed design alterations differentiated the V from its regular-line CTS siblings. All these changes added a just-right dollop of ominousness to the CTS's already aggressive shape.

Cadillac claimed the CTS-V could accelerate from 0–60 mph in 3.9 seconds and do the quarter mile in twelve seconds at 118 mph, but magazine road testers generally couldn't replicate those numbers. Most road tests returned 0–60 times in the low-to-mid fours and quarter miles in the low-to-mid twelves—still world-class. Cadillac engineers aimed for a "bimodal nature" with the CTS-V, so the near-supernatural performance capabilities didn't seriously compromise the luxury and refinement expected of a premium-brand sedan. Outside of a noticeably stiffer ride and a menacing exhaust note, the CTS-V gave up little to its tamer siblings in mundane stop-and-go driving.

Cadillac was eager to prove the CTS-V's supercar bona fides beyond the expected routine of buff-book road tests and "shootouts." In May 2008, GM Performance Division executive John Heinricy piloted a stock automatic-transmission '09 CTS-V to a blistering 7:59.32 lap of Germany's famed Nürburgring Nordschleife road-race course. Cadillac claimed this was the first sub-eight-minute lap of the 'Ring recorded by a production performance sedan on street tires.

For 2011, a slick coupe and station wagon joined the CTS-V sedan. The coupe and wagon were mechanically identical to the sedan but possessed the obvious benefits and drawbacks of their respective body configurations. Compared to the sedan, both had slightly compromised rear visibility. The wagon came standard with a power liftgate and nearly doubled the sedan's carrying capacity, with twenty-five cubic feet of cargo space behind the rear seats. With the rear seats folded, that number grew to 53.4 cubic feet.

Regardless of body style, the second-gen CTS-V's "Standard of the World" levels of performance went a long way toward erasing memories of some embarrassing Cadillacs from the then, not-too-distant past.

ROUSH
427R MUSTANG 2010

Every Roush Mustang 427R started life with good bones, those of the Ford Mustang. Plucked from the assembly line at the peak of freshness, lucky Mustang GTs were selected from the queue destined for dealerships and redirected to Roush Performance in Livonia, Michigan.

Founded by racing jack-of-all-trades and one-time Ford engineer Jack Roush, Roush Performance has been modifying Mustangs since 1998. A 2010 *Consumer Guide®* Automotive Best Buy, the stock Mustang GT was already a fine ride. In the hands of Roush Performance, those cars were morphed in a manner akin to Pinocchio's conversion from wooden puppet to real boy.

And like Pinocchio, the 427R had a good heart. Already beating strongly in the stock Mustang GT, the 4.6-liter three-valve V8 was boosted by a healthy 120 bhp compliments of Roush Charging. A supercharger designed and built in-house by Roush, the intercooled Roush Charger, along with a few other tweaks raised peak engine output to 435 bhp and a stout 400 pound-feet or torque. An upgraded cooling system was also part of the deal.

Predictably, there was more to the 427R package than just power. To help this puppet-turned-boy learn to walk without strings, Roush thoroughly worked over the suspension, and replaced the front and rear springs, struts, and sway bars with equipment from the Roush Performance Catalog. Specific wheels and tires were also included.

Also part of the package were cabin and body-trim enhancements too numerous to list here. Understandably proud of its creation, Roush took care to apply its name to the vehicle more times than we could have accurately counted.

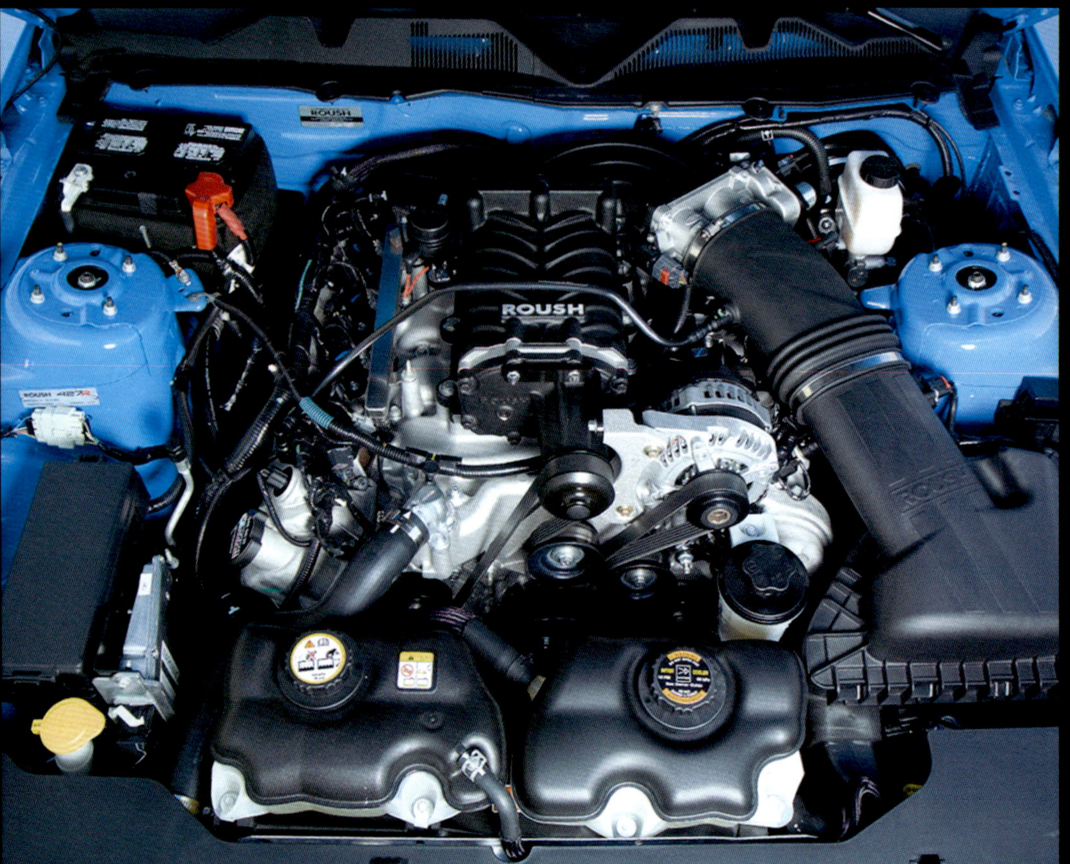

But as interesting as the 427R might have been on a spec sheet, it was on the road that it shined best. According to *Motor Trend*, the 2010 427R would reach 60 mph from a stop in 4.7 seconds. Per *MT*, that was about two tenths of a second faster than a stock manual-transmission Mustang GT. While at first blush those numbers might have disappointed, it was important to note how much wheel spin needed to be overcome to have achieved anything like an efficient launch. All the torque this delightfully revvy engine produced was available immediately, utterly without the expected blower lag or hesitation.

Helping manage all this power was an optional short-throw shifter. This hefty, solid-feeling stick shift was a paragon of mechanical precision and smoothness that invited unneeded shifting simply because it was so much fun to manipulate.

This powertrain responsiveness became an allegory for the entire car. No Mustang, no matter how special or rare, had ever felt as much an extension of the driver as the 427R did.

The steering was direct and surprisingly light. The suspension upgrades resulted in a car that was not only light on its feet and absolutely flingable—in the best possible way—but one that actually rode more comfortably and with more composure than any stock Mustang.

All that said, the power, handling, and ride accounted for maybe half of what made the 427R so special. There was also the sound. The snorting, throbbing, menacing burble emitted from the 427R's exhaust was not only intoxicating, it was addictive, an aural tachometer that demanded unwarranted throttle blips no matter your actual power needs.

TESLA MODEL S 2012

While established car companies seemingly sat on the sidelines, convinced that electric vehicles (EVs) had to be small, affordable, and practical, California-based Tesla reached wealthy early adopters with a car that was good looking, exciting to drive, and packed with counterculture cachet. Properly equipped, a Model S would travel 300 miles on a charge and zipped to 60 mph in three seconds. The Model S changed countless minds regarding the potential of battery-powered vehicles and set the stage for other makes to enter the electric-car field.

MORGAN

3 WHEELER 2012

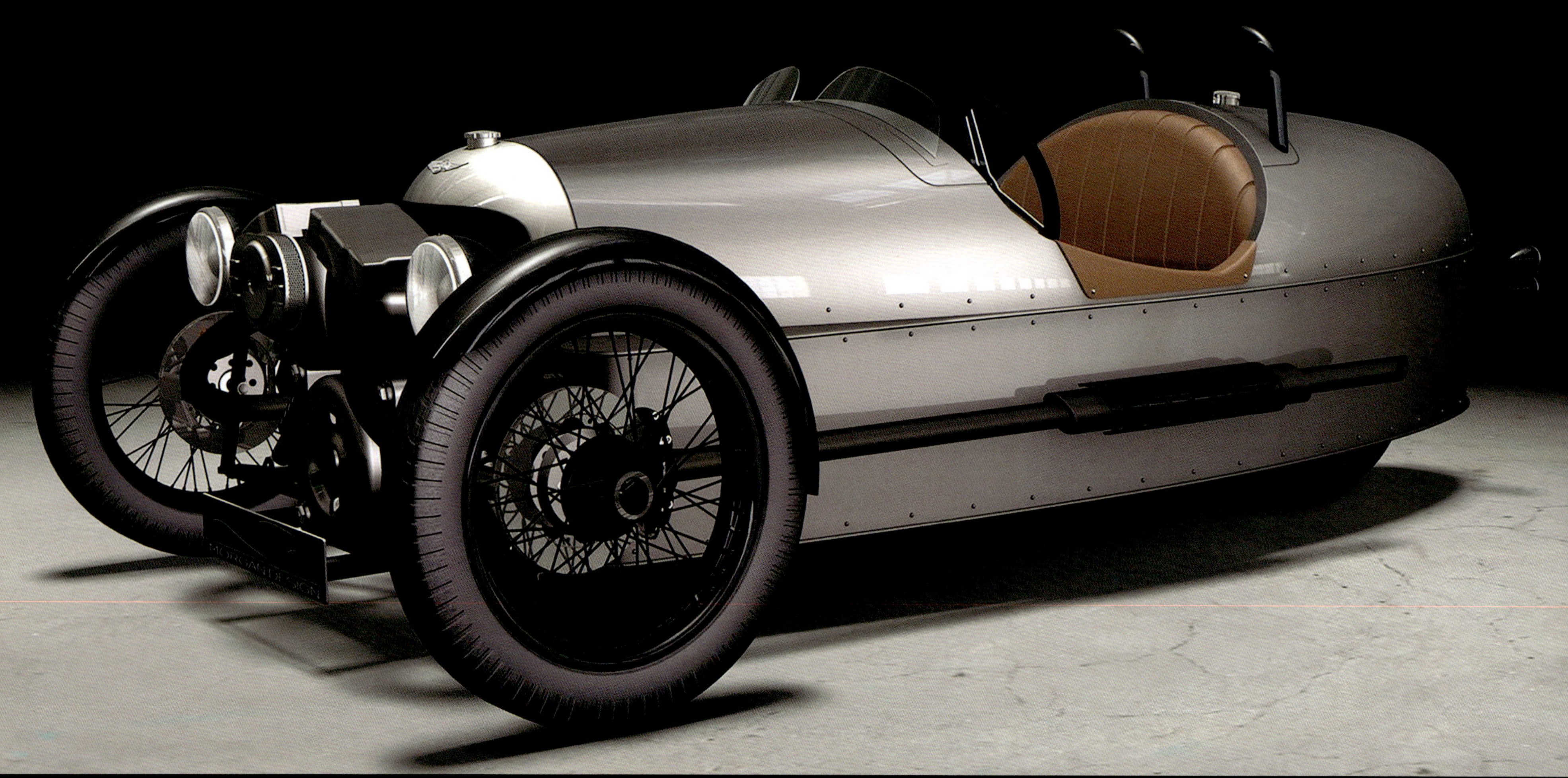

A product of storied British "sports car" maker Morgan, the quirky 3-Wheeler is one of the most aptly named vehicles ever sold in the U.S. The tiny roadster seats just two, and it's best if those two are, let's say, on the slender side.

The 3-Wheeler is powered by a 2.0-liter V-twin motorcycle engine mounted in front of the vehicle. A 5-speed manual transmission directs power to the single rear wheel. Weighing just 1200 pounds, the 80 horsepower engine is capable of pushing the tiny sports car to 60 mph in just 4.5 seconds.

MORGANDESIGN

CHEVROLET

SS 2014-2016

When discussing the 2014-16 Chevrolet SS, it was impossible to ignore the one-year-and-out 2009 Pontiac G8 GXP. Killed off like all Pontiacs in the wake of General Motors's 2009 bankruptcy filing, the 415-bhp GXP was the highest-performance variant of Pontiac's new-for-2008 G8 based on GM's Holden Commodore VE from Australia.

Commodore production continued with versions of the car sold as Chevrolets in international markets. Around the time that Holden confirmed an updated Commodore VF model in 2013, GM issued word that a variant would come to America as the 2014 Chevrolet SS.

In the States, Chevrolet had been selling a long-wheelbase Commodore spin-off as the police-only Caprice PPV since 2011. However, like the departed G8, the new SS was based on the standard-length rear-wheel-drive Commodore. Thus it rested on a 114.8-inch wheelbase and was 195.5 inches long.

Americans were only offered well-equipped SS sedans priced from $44,470. The lone powerteam was a 415-horsepower 6.2-liter V8 and six-speed automatic transmission. The suspension paired MacPherson struts up front with an independent multilink set-up out back. Brembo front brakes and forged 19-inch alloy wheels wearing sticky Bridgestone rubber were part of the deal too.

Exterior styling closely followed the updated Commodore. High-intensity-discharge headlamps and LED daytime running lights were standard. The hood and trunklid were aluminum to save a bit of weight. The nine-inch-wide rear wheels were .5-inch wider than the fronts, and wore bigger rubber. Five colors were available: Silver Ice Metallic, Red Hot 2, Phantom Black Metallic, Heron White, and Mystic Green Metallic.

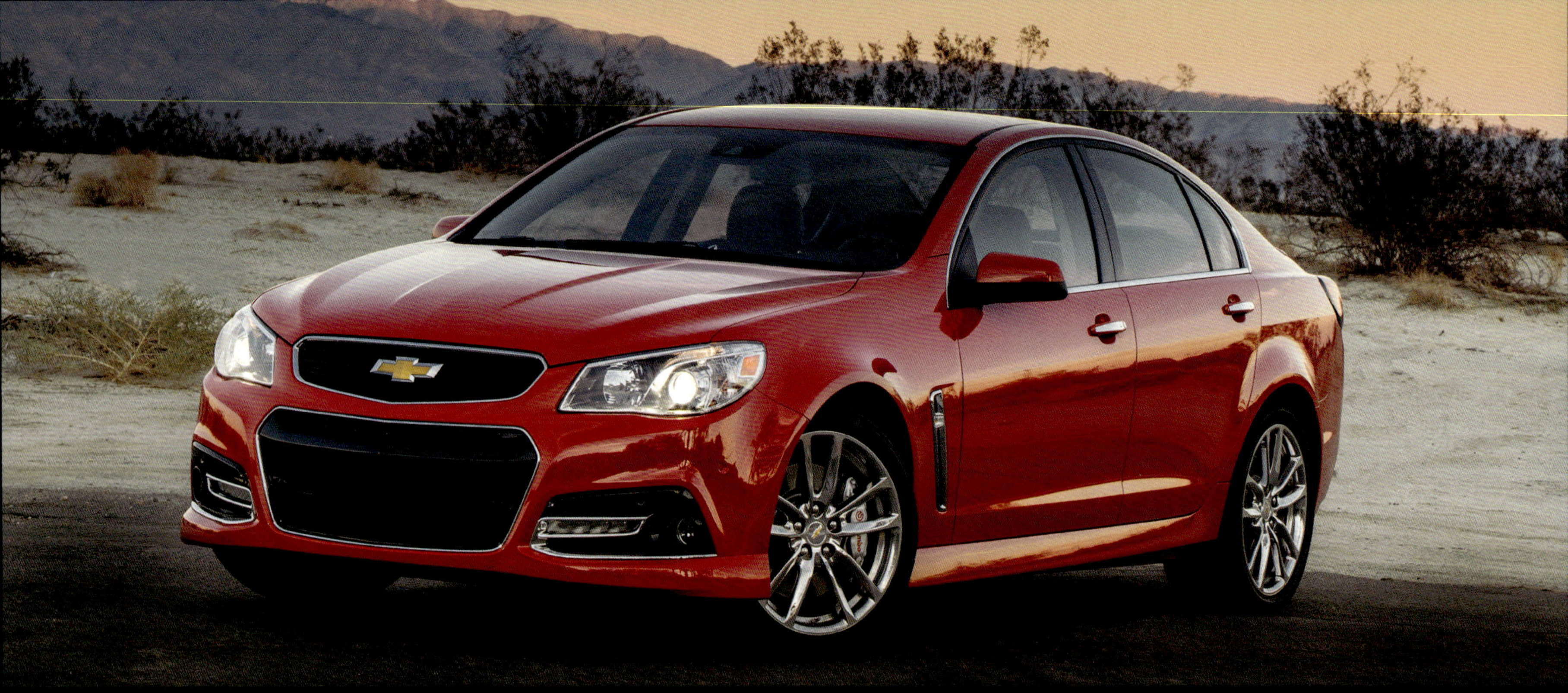

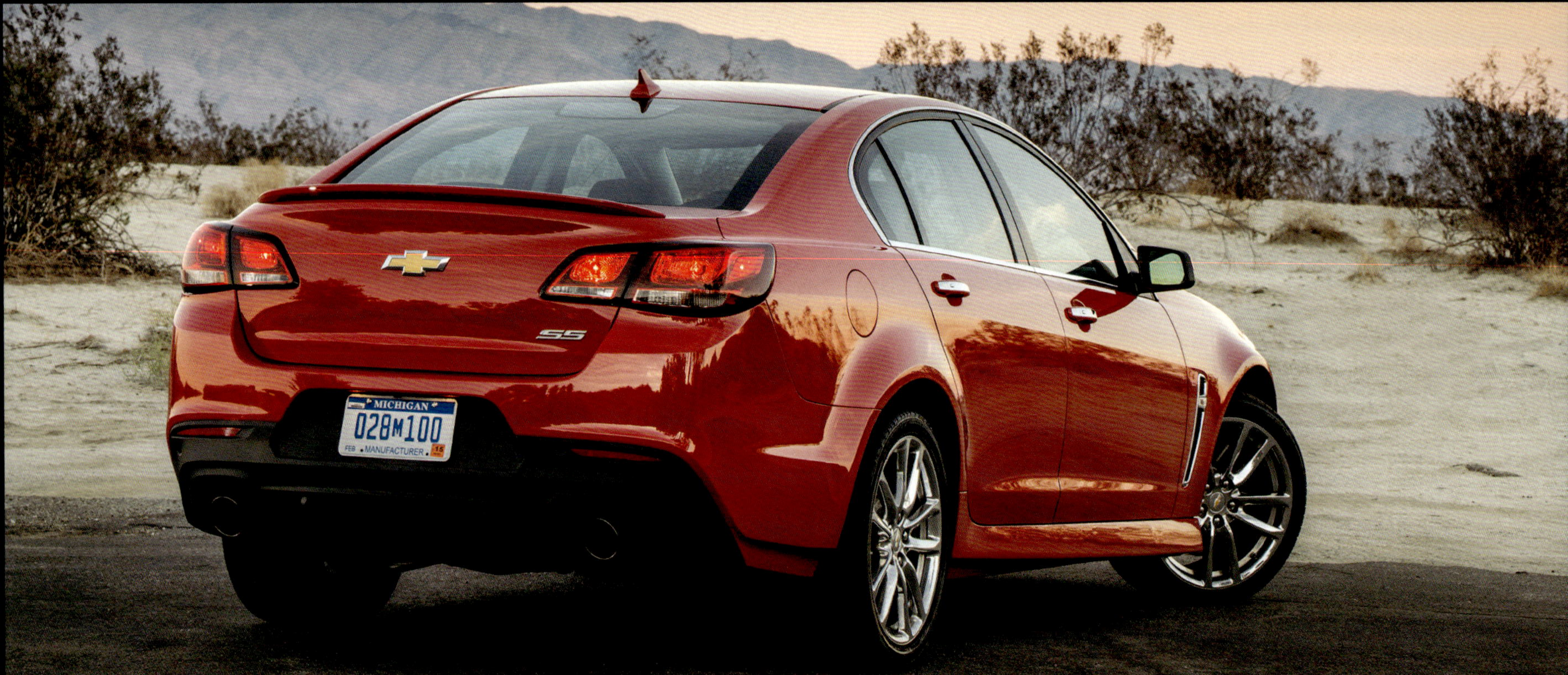
MICHIGAN
028M100
MANUFACTURER

All interiors had black leather upholstery and eight-way power front buckets. A Bose stereo was standard, along with Chevy's MyLink infotainment and touch-screen navigation. A power sunroof was one of the car's few options.

Road & Track tested an SS for its February 2014 issue, reporting a 0-60-mph time of 4.5 seconds and a quarter-mile dash of 12.9 seconds at better than 110 mph. *R&T* liked the balanced handling, and said the car "feels like an American version of the BMW M5." Gripes centered on the body's reserved styling and chrome accents. The 2015 SS received GM's Magnetic Ride Control suspension and rear Brembo brakes as standard. Perhaps the biggest news was that a six-speed manual transmission packaged with a more aggressive 3.70:1 axle ratio joined the order sheet. New metallic colors were on an expanded palette, including Perfect Blue, Some Like it Hot Red, Alchemy Purple, Jungle Green, and Regal Peacock Green.

Chevy gave the 2016 SS a subtle styling update with a new front fascia and functional hood vents. The 19 inch wheels were cast rather than forged. There was also a new dual-mode exhaust system that allowed a throatier sound at full throttle. Slipstream Blue Metallic paint was in; Perfect Blue and Alchemy Purple were dropped.

GM announced that it would stop making cars in Australia by the end of 2017. In September 2015, Australian outlet *Cars-Guide* reported that GM Asia-Pacific boss Stefan Jacoby confirmed the Chevrolet SS would be phased out and not directly replaced when production of the V8 powered Commodore ended.

BMW I8 2014-2019

Beginning with the 2009 introduction of its Vision Efficient Dynamics concept vehicle, BMW began teasing the idea of an exotic sports car with a high-tech, eco-conscious focus. After exploring the concept further (and introducing the i8 name) on a couple subsequent concept vehicles, the company officially committed to a production model—in September 2013, the i8 prototype was unveiled at the International Motor Show in Frankfurt, Germany.

The production i8 launched as a 2014 model, with unorthodox styling that covered an equally unconventional plug-in-hybrid powertrain. And almost all of the concept vehicles' outlandish features—most notably the scissor-wing doors—made the jump from the show floor to the showroom. The starting price was steep ($135,700 in the United States), but not unreasonable compared to similar luxury exotics.

The i8's gas-electric powertrain paired a turbocharged 1.5-liter 3-cylinder gasoline engine rated at 228 horsepower with a BMW-made synchronous electric motor that was good for 129 hp. Combined maximum output was 357 horsepower and 420 pound-feet of torque. The gasoline engine drove the rear wheels and was mated to a six-speed automatic transmission; the electric motor sent its power to the car's front wheels via a two-stage automatic transmission. Power for the electric motor was supplied by a liquid-cooled lithium-ion battery pack with a usable capacity of five kilowatt hours. The powertrain control software allowed the car to be operated on either power source independently or use both of them together, depending on the driving situation; the front-to-rear power split was also variable. BMW quoted the i8's 0–60 mph time at 4.2 seconds, and the EPA's gas-electric mileage estimate was 76 MPGe in combined city/highway driving.

Along with the smaller BMW i3 commuter vehicle, the i8 utilized BMW's LifeDrive architecture concept, which utilized independent "Life" and "Drive" structural modules. In the case of the i8, the Drive module was an aluminum structure that housed the gas and electric motors, the lithium-ion battery pack, chassis and suspension hardware, and the car's crash structure. The Life module consisted of the car's CFRP (carbon fiber reinforced plastic) passenger cabin. The i8 was 184.6 inches long, 76.5 inches wide, and 51 inches tall, with a curb weight of 3285 pounds.

The interior used a sports-car-typical low seating position. The cabin's appearance followed the layered approach used on the body, along with the imaginative use of contrasting colors. A mix of leathers, cloth accents, painted surfaces, and exposed carbon fiber added to the ambiance. The driver could choose from five driving modes using the Driving Experience Control switch and eDrive button.

Consumer Guide® Automotive editors tested a 2015 i8, and were impressed by the car's balance of ride comfort and sporty handling, as well as its vivacious acceleration and throttle response. However, the swoopy, low-slung styling and radical doors didn't do any favors for ease of entry and exit or visibility.

BMW announced several updates for the 2019 i8, including the addition of a two-seat roadster variant to be sold alongside the coupe. Other upgrades included an increase of total gas/electric output to 369 horsepower, and an updated lithium-ion battery pack that increased capacity (BMW said net capacity was 9.4 kilowatt hours) and range. There were also some new interior trim choices and exterior colors. Coupe prices started at $147,500, and the new Roadster model had a base price of $163,300.

DAIHATSU

COPEN 2015

To avoid paying certain taxes, Japanese consumers have the option of purchasing a "kei" car, a class of tiny vehicles that are exempt from many local and federal fees. Though small, many kei cars are surprisingly fun to drive. One such example was the Copen. Built by Toyota subsidiary Daihatsu, the diminutive front-wheel-drive convertible was powered by a 658-cc 3-cylinder engine rated at 67 horsepower. With just 1900 pounds to move around, the little turbocharged engine provided brisk acceleration. Available only with the steering wheel located on the right-hand side of the car, the Copen was never exported to the U.S.

HUM3D

CHEVROLET

CORVETTE Z06 2015

The C7-generation Chevrolet Corvette that arrived in 2014 was significantly lighter and stiffer than the C6 Corvette it replaced. A new "King of the Hill" version came one year later with the revival of the super-performance Z06 model. The C7 Z06 had a supercharged 6.2-liter V8 with 650 horsepower and 650 pound-feet of torque, which enabled it to accelerate from 0-60 mph in 3.2 seconds with the 7-speed manual transmission, or a jaw-dropping 2.95 seconds with the 8-speed automatic. *Road & Track* achieved a top speed of 186 mph. In spite of its incredible performance, the Z06 was surprisingly livable in everyday driving. Plus, the base price began at under

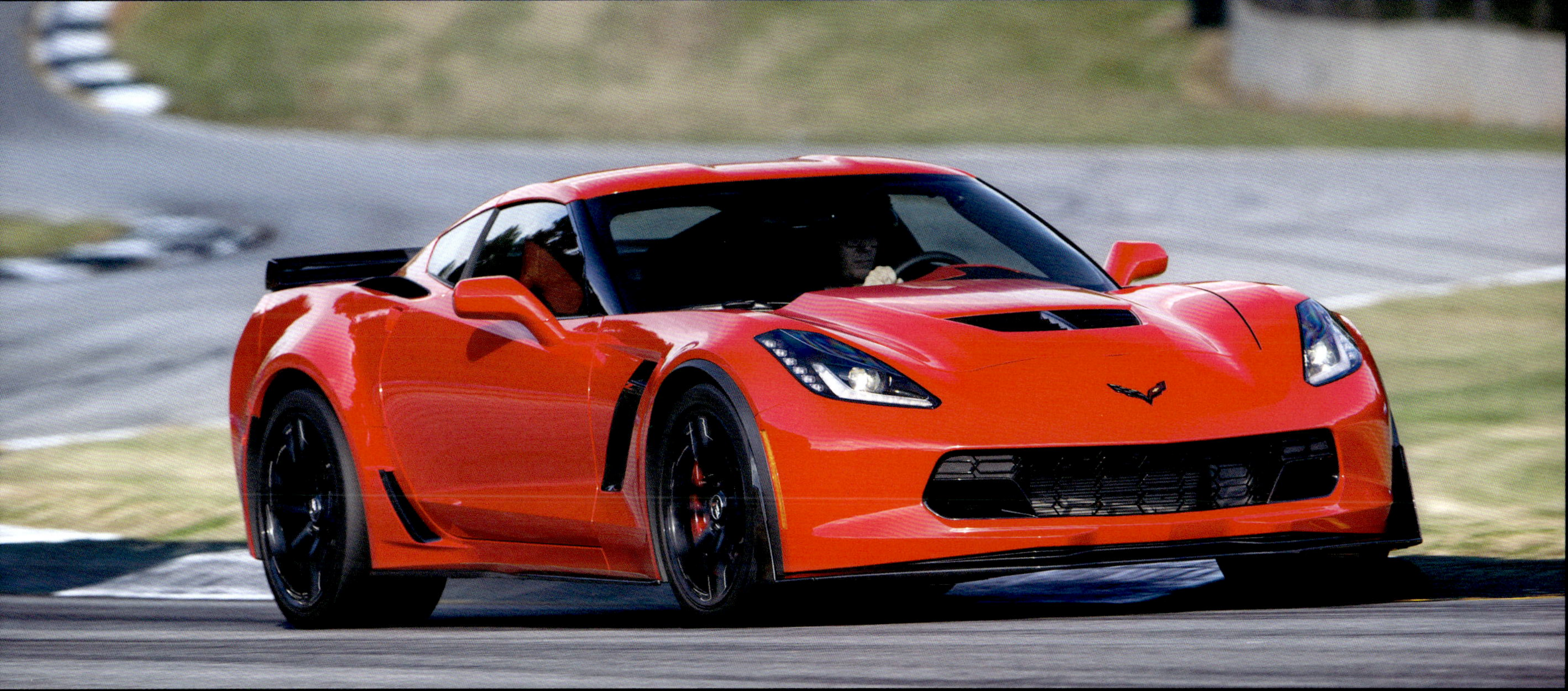

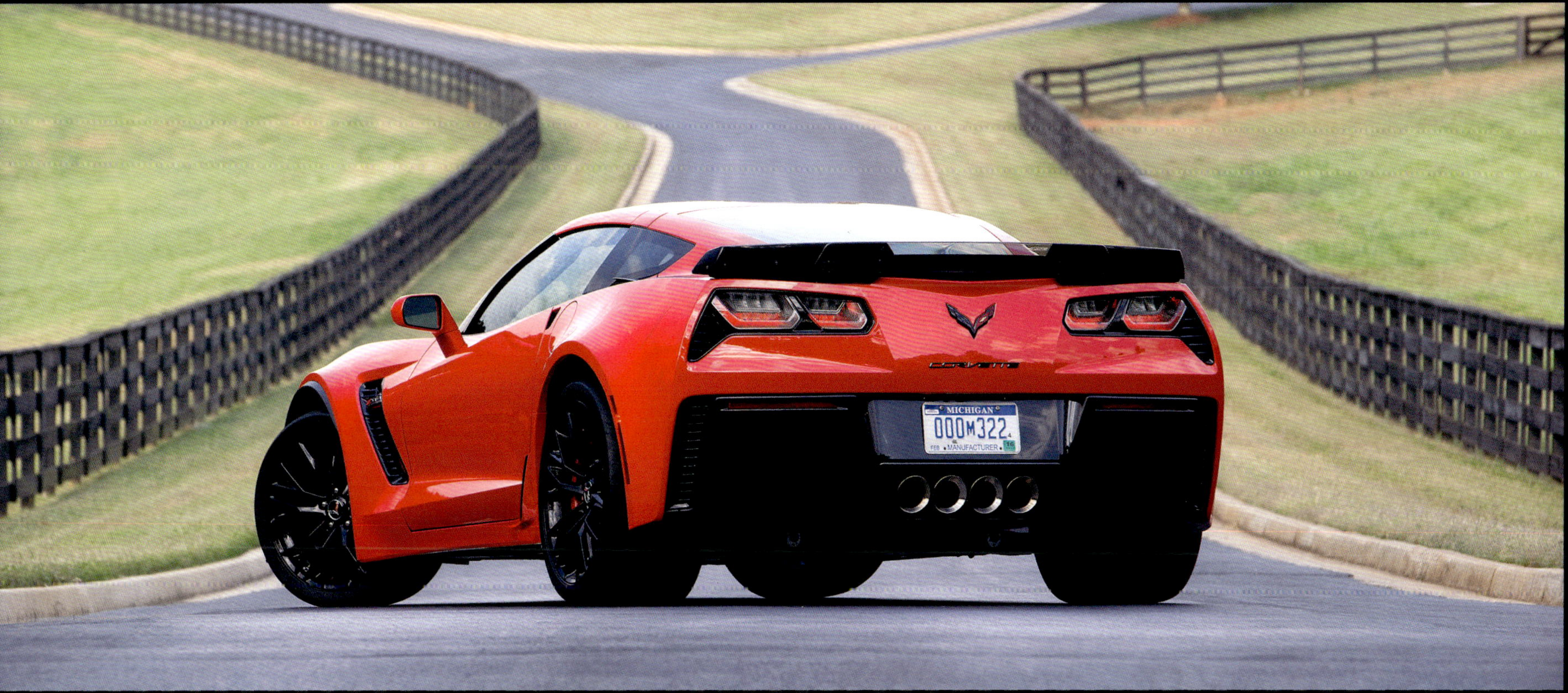
MICHIGAN
MANUFACTURER

ALFA ROMEO

4C 2015-2016

Alfa Romeo's long-promised return to the American market had seemingly been in the works since imports of the 164 sedan had ended in 1995. Maserati dealers sold Americans about 100 $250,000-plus Alfa 8Cs between 2008 and 2010. But more attainable models were only promised, not delivered. Finally, after countless delays, the first examples of the Alfa Romeo 4C hit the company's small network of Stateside dealers in late 2014.

Not exactly mainstream, the 4C was a small, two-seat, mid-engine sports car that was built at corporate cousin Maserati's plant in Modena, Italy. The main section of the chassis is a tub made of carbon fiber that remained visible in the car's interior. Roof reinforcements and a frame that mounts the engine were aluminum, as were the front and rear chassis structures. The bodywork was formed from sheet-molding compound, a plastic composite. The 4C rode on a 93.7-inch wheelbase and was 157 inches long.

The 4C's engine was transversely mounted just forward of the rear wheels. It was an all-aluminum turbocharged 1.7-liter four-cylinder job rated at 237 bhp and 258 pound-feet of torque. It mated to a six-speed dual-clutch automated transmission that had paddle shifters for manual gear selection. The Alfa "DNA" selector allowed the driver to choose between several modes that altered engine, transmission, and stability-control calibration.

The horsepower number was modest by today's standards, but the coupe's claimed 2,465-pound curb weight helped. *Road & Track* reported a 0–60 mph time of 4.2 seconds and a quarter-mile run of 12.8 seconds at 108.2 mph.

The first of the American-market 4Cs were Launch Edition coupes in a run of just 500 cars. These had additional interior and exterior carbon fiber accent pieces, black microfiber and leather interior trim with red or white contrast-stitch detailing, side air intakes on the front fascia, and a serial-numbered plaque. Performance enhancements included "race" exhaust and suspension tuning, along with 18-inch front and 19-inch rear wheels and tires. Base price was $68,400.

A "base" 4C coupe came later with smaller 17- and 18-inch rubber, less equipment, and a $53,900 starting price. The 2015 lineup also included a $63,900 4C Spider with a removable reinforced-fabric roof panel. The 2,487-pound Spider also introduced interior updates that included additional small-item storage spots and an Alpine-brand audio system. These appreciated tweaks helped mitigate complaints about the early cars, especially the roundly panned Parrot-brand stereo.

For 2016, the 4C coupe received the Spider's interior updates and Alpine head unit. There were new interior trim choices for both cars too.

For the hard-core sports-car lover, this exotic Italian was relatively affordable. The car's surprising mixture of high-tech materials and back-to-basics personality offered an undeniably compelling character as well.

ALFA ROMEO

MICHIGAN
033M005
MANUFACTURER

4C

HONDA S660 2016

The S660 answered the question, "What if Ferrari built a tiny affordable car?" Honda's home-market-only convertible featured a mid-engine design combined with rear-wheel drive and a curb weight comfortably under 1900 pounds.

Power came from a turbocharged 3-cylinder powerplant which mated to either a 6-speed manual or CVT-style automatic transmission. To the delight of Japanese driving enthusiasts, the S660 started at just $18,000. The bad news? Once the S660's soft top was folded and stored under the car's hood, there was absolutely no room for additional cargo.

S660

S660

S660

LAMBORGHINI

AVENTADOR S 2016

Ferruccio Lamborghini chose a charging bull as the emblem for his car company, and had a tradition of naming models after famous fighting bulls. That practice continued under the Lamborghini firm's various corporate owners over the years—the Aventador, which debuted as a 2011 model, was named for a fighting bull from the nineties. For 2016, an Aventador S version was introduced; it featured a more sophisticated suspension and a four-wheel steering system that allowed sharper steering at low speeds and better stability at high speeds. There was also a more powerful 740 horsepower 6.5-liter V12 that could push the all-wheel drive, mid-engine supercar to 217 mph, with 0-62 mph coming in just 2.9 seconds. The S's base price was around $450,000.

LB 734CS

Lamborghini

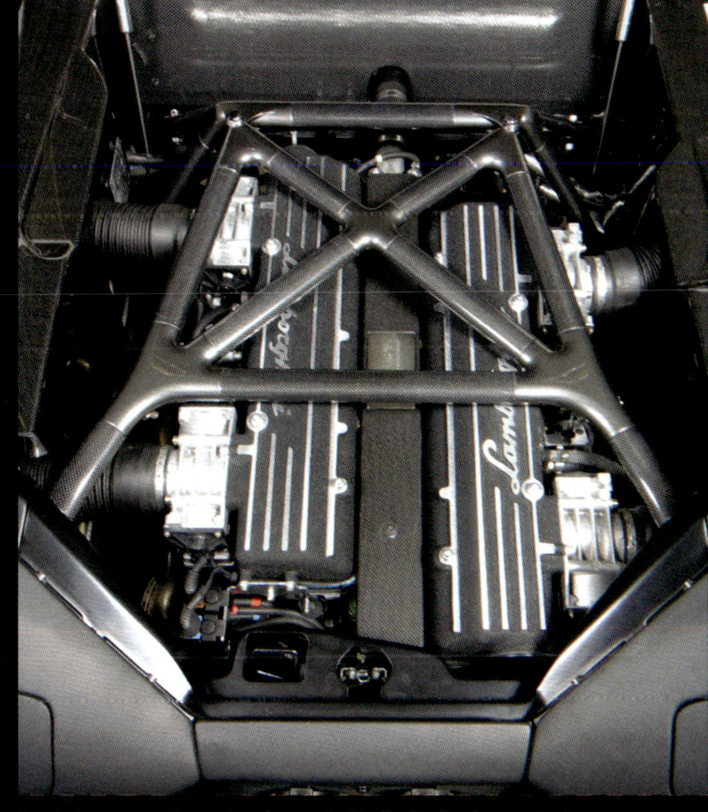

BMW M2 2016-2017

In 2014 BMW replaced the 1-Series as its premium subcompact in America. The new holder of that job was the 2-Series. The basic idea here was a coupe or convertible with four seats and rear drive. There were two subseries, the 228i with a turbocharged four and the M235i with a 320 bhp turbocharged inline six.

What was missing at the time was a full performance version from BMW's M division. That gap was filled in 2016 when the Bavarians brought out the M2 coupe.

To make it, the M235i's 3.0-liter six was reworked and received the pistons and crankshaft main-bearing shells from the engine used in the larger M3 and M4. The oil supply was tweaked to ensure proper lubrication under heavy acceleration and braking, and the cooling system was upgraded. Fitted with a single twin-scroll turbocharger, the M2's heart was rated at 365 bhp.

The standard transmission was a six-speed manual. It came with an automatic rev-matching feature that electronically blipped the throttle for downshifts. (Some buff-book reviewers bemoaned that rev matching could only be disabled by turning off the car's stability-control system.) A seven-speed dual-clutch transmission—"M DCT with Drivelogic" in BMW-speak—was optional. This gearbox allowed for automatic operation, or the driver could shift gears manually without using a clutch pedal. BMW quoted a 0-60-mph time of 4.2 seconds with the DCT, and 4.4 with the traditional stick.

M2's disc brakes and aluminum-intensive front and rear suspensions were borrowed from the M3 and M4. The 19-inch forged wheels were nine inches wide in front, 10 inches wide in the rear, and wore Michelin Pilot Super Sport rubber.

The M2 bowed at 176.2-inches long, rode a 106-inch wheelbase, and had a curb weight of 3,450 pounds with manual transmission. For comparison, a 2017 Chevrolet Camaro was 188.3 inches from nose to tail on a 110.7-inch wheelbase, and weighed 3,463 pounds with a V6 and manual trans.

M2 bodywork deviated substantially from the standard 2-Series cars. Front and rear fenders were dramatically bulged to accommodate the car's wider tracks and sticky Michelins. The car also wore model-specific front and rear fascias, along with a modest lip spoiler on the trunklid.

BMW cataloged few options for the M2. Alpine White was the only "free" paint choice. Alternatives were the extra-cost metallic Long Beach Blue, Black Sapphire, or Mineral Grey. Interiors were trimmed in black leather with contrasting blue stitching. Accents included "open-pore carbon fiber" trim and some Alcantara-covered bits. There was also an Executive Package that added a heated steering wheel, rearview camera, and some driver-assistance items.

The introductory starting price of $51,700 jumped to $52,695 for 2017. Changes for '17 were the additions of wireless smartphone charging and a Wi-Fi hotspot to the Executive Package. There was a new $2,500 M Driver's Package that increased top speed to 168 mph (from 155) and included a driver-training course at a BMW Performance Center in South Carolina or California.

ACURA NSX 2017

Acura first dipped a toe into the mid-engine exotic-car market for 1991, when it introduced the NSX—a two-seat, mid-engine supercar powered by a naturally aspirated V6. The first-generation NSX was a pragmatic exotic, offering outstanding reliability and ergonomics, but perhaps not inspiring the same level of irrational passion as its European rivals. Production of the first NSX lasted from 1991 to 2005, and no major changes were made over the model's lifespan. After a long gestation period, a reborn NSX finally debuted for 2017. The second-generation model was an all-wheel drive hybrid with a twin-turbo 3.5-liter V6 and three electric motors. One motor was integrated with a 9-speed dual-clutch automatic transmission and joined the gas engine in powering the rear wheels; the other two motors drove the front wheels. The combined output of all motors was 573 horsepower. According to *Road & Track*, the rakish hybrid was capable of accelerating 0-60 mph in 3.1 seconds and had an estimated top speed of 191 mph. The NSX was an exotic that was refined, reliable, and had the added bonus of respectable fuel economy from its state-of-the-art hybrid drivetrain.

FERRARI

LAFERRARI 2017

Ferrari would probably be the last make expected to sell a hybrid, but the LaFerrari was no ordinary hybrid—it combined a 788 horsepower V12 with a 161 hp electric motor for a total 949 hp. Introduced for 2013, the LaFerrari hybrid was even faster than Ferrari's celebrated 2003-04 Enzo model—it could rocket from 0-60 mph in less than three seconds and hit a top speed of 217 mph. Plus, the weight of the low-mounted battery packs lowered the car's center of gravity and improved handling. The LaFerrari's price tag was a cool $1.4 million, but Ferrari had no trouble selling 500 coupes, as well as 210 Aperta convertible versions. The final LaFerrari Aperta was sold at an auction in 2017 for $9.96 million, with proceeds supporting the Save the Children charity.

BUGATTI CHIRON 2017

There are supercars, and then there are supercars, and the Bugatti Chiron definitely qualifies as the latter. The Chiron's quad-turbo-charged 8.0-liter W-16 engine developed an astonishing 1500 horsepower and 1180 pound-feet of torque—enough for a 0-60 mph time of 2.4 seconds and an electronically limited top speed of 261 mph. A Haldex all-wheel drive system got all that power to the road with the least amount of drama. The Chiron set a world record by accelerating 0 to 400 km/h (248 mph) and then braking to a stop in 41.96 seconds. That record was quickly bettered by Koenigsegg, but it was still a remarkable accomplishment. Base price for the Chiron was around $3,000,000. As expected at that price, there was no visible plastic in the cockpit—only top-grade leather and metal. The Chiron was more refined and quieter than its predecessor, the Bugatti Veyron, and was also surprisingly docile at low speeds.

FORD GT 2017

To celebrate the 50th anniversary of the Ford GT40's historic 1966 Le Mans victory, Ford designed an all-new GT racecar for the 2016 24 Hours of Le Mans. The new GTs did their legendary forebears proud, finishing first, third, fourth, and ninth in the GTE Pro Class. The production GT was closely related to the racing version, and lacked the luxury features often found on other high-end supercars. The GT's cockpit was a tight fit for two passengers, and cargo room was almost nonexistent. The payoff was reduced weight, with racecar-like performance and handling. The EcoBoost 3.5-liter V6 shared its basic engine block with the Ford F-150, but developed 647 horsepower and was capable of traveling from 0-60 mph in 2.9 seconds and reaching a top speed of 216 mph, according to *Car and Driver*. The price was around $450,000.

MICHIGAN
039M493

CHEVROLET

CAMARO ZL1 2017-2018

Chevrolet celebrated the Camaro's 50th anniversary in 2017. There was a commemorative model with special trim, but more importantly there was a high-performance ZL1. The original Camaro ZL1 dates to 1969. Available by special order through Chevy's Central Office Production Order system, the first ZL1 was a dragstrip terror powered by a shockingly expensive all-aluminum 427-cid "big-block" V8. Only sixty-nine were produced. Modern-day ZL1 history began in 2012, when the name was revived for an all-around performance Camaro powered by a supercharged 580 bhp 6.2-liter V8.

Camaro was redesigned for 2016. The sixth-generation model closely followed the appearance of the well-received fifth-gen car, and was based on the General Motors "Alpha" rear-drive platform shared with Cadillac's ATS and CTS. Perhaps most significantly, the new Camaro was lighter and marginally smaller than the car it replaced.

The latest ZL1 debuted for 2017 in coupe and convertible forms. Unique exterior styling touches included wider front fenders, a vented hood, larger front splitter, and a rear wing. Power came from the LT4 supercharged 6.2-liter V8 good for 650 horsepower and 650 pound-feet of torque. It could be backed by a six-speed manual transmission with an active rev-matching function or a Hydra-Matic ten-speed automatic. Other upgrades included a specifically tuned "Magnetic Ride" adaptable suspension, Brembo-brand brakes, 20-inch forged aluminum wheels, and ZL1-specific Goodyear Eagle F1 Supercar rubber. Weight was down 200 pounds compared to the previous ZL1.

Chevy-supplied performance numbers were a 0–60 mph time of 3.5 seconds, a quarter-mile sprint of 11.4 seconds at 127 mph, and a cornering limit of 1.02g. *Motor Trend* duplicated Chevy's 0–60 mark, but the best it could do in the quarter was 11.5 seconds at 125 mph. Still, that's .2 quicker than MT coaxed out of a Dodge Challenger Hellcat. *Car and Driver* said, "It's a car you can live with every day and hustle across any piece of pavement, and we wouldn't change a thing."

For 2018, the ZL1 added a new coupe-only 1LE variant focused on racetrack performance. It added specific aerodynamic features to increase downforce, including air deflectors and dive planes up front and a carbon fiber rear wing. The suspension was upgraded with what Chevrolet called "Multimatic DSSV" (Dynamic Suspension Spool Valve) dampers, adjustable front ride height and camber, and a three-way-adjustable rear stabilizer bar. The 1LE also wore wider, and lighter, 19-inch forged wheels and specially designed Goodyear Eagle F1 Supercar 3R tires. Chevy said the 3,820-pound curb weight is about sixty pounds lighter than a base ZL1 coupe thanks to the suspension and wheel changes, thinner rear glass, and a fixed rear seat back.

Visual cues marking the 1LE included a satin-black hood, black side mirrors and wheels, and darkened taillamps. The LT4 engine was unchanged, but 1LE only came with a six-speed manual gearbox running a model-specific top-gear ratio. Chevy asserted the ZL1 1LE was three seconds a lap faster than a standard ZL1 coupe when tested at GM's Milford Road Course test track. The 2017 ZL1 coupe was priced from $62,135, with the convertible starting at $69,135. The 2018s were up to $63,795 and $69,795, respectively. The 1LE package added $7,500 to the coupe's bottom line.

For Camaro's 50th anniversary, Chevy's best present to Camaro fans was the ZL1. It was crazy fast in a straight line or on a road course, while still impressing the buff-books on the street.

DODGE DEMON 2018

The 2018 Dodge Challenger SRT Demon wasn't a supercar in the conventional sense, especially since it offered so much performance for the relatively reasonable starting price of $86,000. Supercars typically seat only two passengers. The Demon could seat up to five, but only if properly optioned—both the rear seat and front passenger seat were deleted for weight savings, but could be added back by the buyer for $1 each. The Demon might have been notably larger than a "true" supercar, but it packed a supercharged 6.2-liter V8 that pumped out 840 horsepower when properly equipped—enough to move the big, hefty coupe with supercar performance, albeit in a straight line. Dodge claimed the Demon was the world's fastest quarter-mile production car, with an elapsed time of 9.65 seconds at 140 mph. Zero to 60 mph was achieved in 2.3 seconds. Top speed was electronically limited to 168 mph, because the tires were essentially street-legal drag slicks that would have been dangerous at the Demon's true top speed. The drag strip, not Nürburgring, was where the Demon excelled.

SUPERCHARGED HEMI

DODGE
DODGE

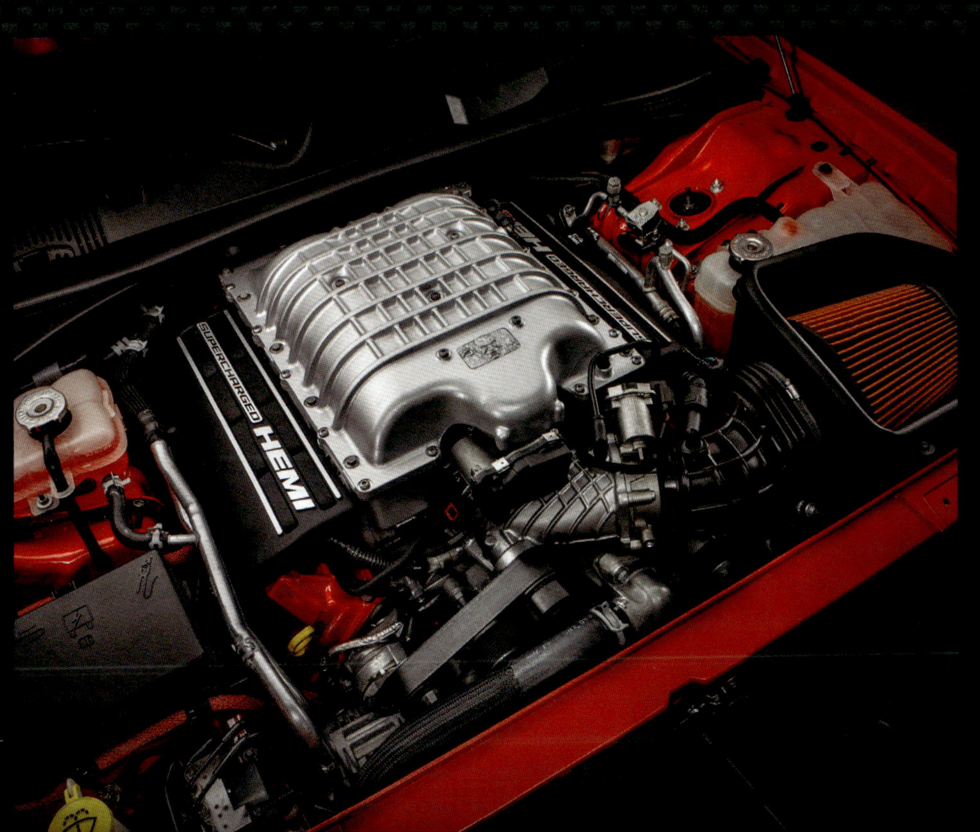

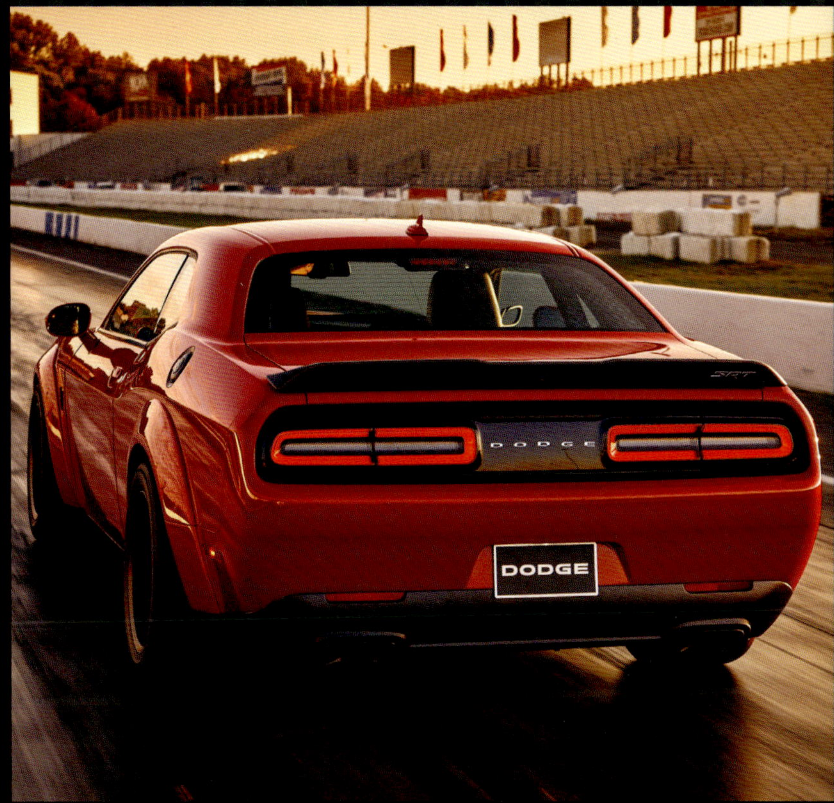

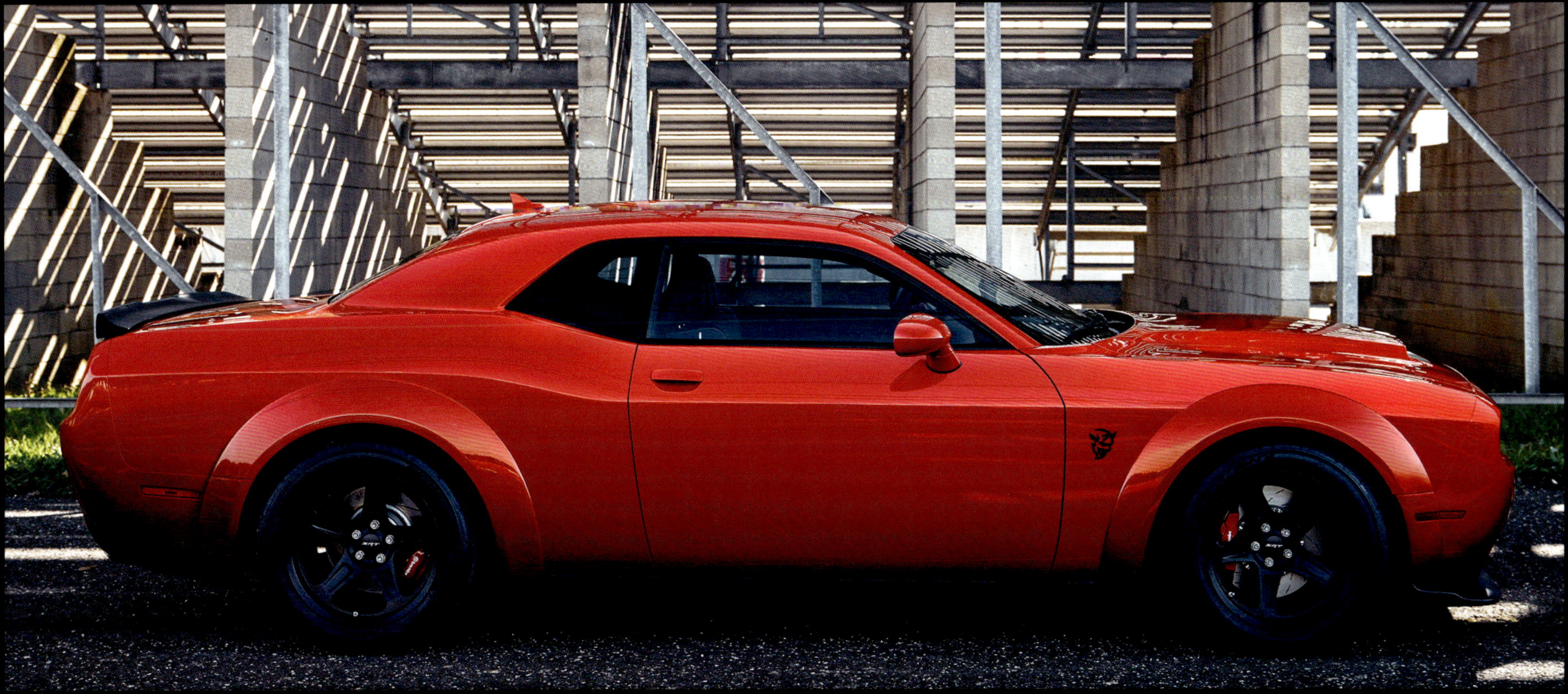

KOENIGSEGG

AGERA RS 2018

Koenigsegg is a small, independent manufacturer based in Sweden that started building cars in the mid-nineties. The Koenigsegg Agera RS turned heads with an official time of 277.9 mph for a two-way average on an 11-mile stretch of highway in Nevada. Actually, the Agera RS's best time was 284.6 mph, but official testing requires an average of travel in both directions. Previously, Koenigsegg broke a record set by Bugatti by accelerating 0 to 400 km/h (248 mph) and then braking to a stop in 36.44 seconds. The Agera RS was powered by a twin-turbocharged 5.0-liter V8 with 1341 horsepower and 1011 pound-feet of torque, and cost in the neighborhood of $2 million.

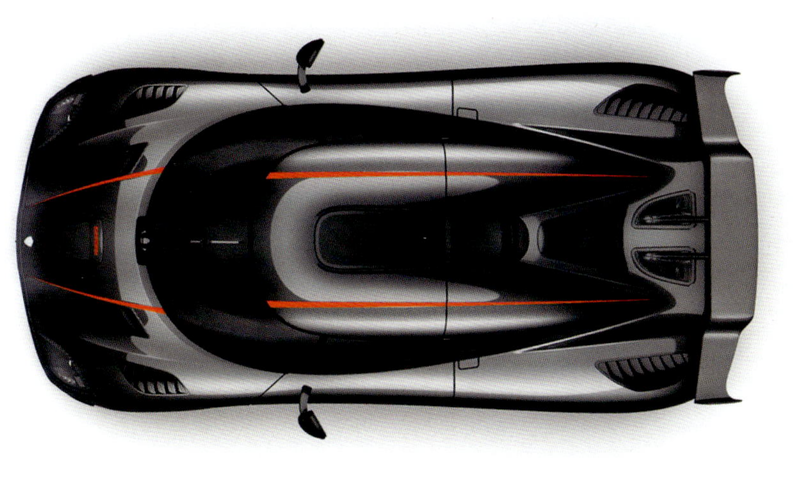

Koenigsegg

ASTON MARTIN

VALKYRIE 2018

Aston Martin partnered with Red Bull Racing of Formula 1 fame to develop an all-new hypercar called the Valkyrie. When first revealed in July 2016, the car was referred to by its codename, AM-RB 001 ("AM" for Aston Martin, and "RB for Red Bull). In March 2017, the Valkyrie name—inspired by the Norse mythological figure—was announced.

The Valkyrie was designed by Red Bull Racing's Chief Technical Officer, Adrian Newey. The form-follows-function exterior styling is a result of Newey's focus on downforce and aerodynamic efficiency; Aston Martin claimed the car "boasts truly radical aerodynamics for unprecedented levels of downforce in a road-legal car," and said that much of the car's aerodynamic downforce was generated by the design of the underside Venturi tunnels.

Noteworthy styling details included exposed headlight assemblies and mounting brackets, a chemically etched aluminum Aston Martin badge on the car's nose that was thinner than a human hair and buried under a coat of clear lacquer, and what was claimed to be the world's smallest center-mount brake light.

The interior filled the space between the car's underbody aero tunnels. Modern-day racecar design provided inspiration, and the cockpit's ergonomics were decidedly minimalist. The seats were mounted directly to the chassis tub, with driver and passenger seated in a reclined position with their legs outstretched in front of them. All switchgear was located on the removable steering wheel, and the car's virtual instrumentation was presented on a lone display screen. Traditional rear-view mirrors were replaced with rear-facing cameras and small monitor screens mounted at each end of the dash.

At the Geneva Motor Show in March 2018, Aston Martin and Red Bull introduced the track-only Valkyrie AMR Pro. Weighing in at 1,000 kilograms (about 2,200 pounds), the car featured numerous weight-saving measures, including lighter-weight carbon fiber bodywork, polycarbonate windows, carbon fiber suspension wishbones, race seats, and a lighter exhaust system. In addition, interior comfort and convenience pieces were deleted.

The 6.5-liter naturally aspirated V12 engine and its hybrid-electric "Energy Recovery System" were tweaked to provide 1,100 horsepower. Top track speed was a claimed 225 mph, and the car was said to be capable of lateral acceleration in excess of 3Gs. Deliveries were expected to begin in 2020, but the company said that the twenty-five examples they planned to build were already sold.

LEXUS LC 500 2018

Lexus called its LC 500 coupe the luxury brand's flagship. It also said it's a concept car come to life, and in this case it's the LF-LC concept first shown at Detroit's North American International Auto Show in early 2012.

The two ideas certainly went hand in hand. The first thing most people noticed about the LC 500 was its expressive styling. The classic long-hood/ short-deck coupe proportions were present, and the racy roofline tapered rearward. Muscular fenders dramatically bulged outward, and bodyside scoops provided an additional sporting touch. Chrome moldings framed the sides of the standard glass or optional carbon fiber roof panel. Other highlights included elaborately-detailed headlight and taillight assemblies and dramatic pop-out door handles.

The LC 500 was built on Lexus' GA-L ("Global Architecture-Luxury") platform that also underpinned the 2018 Lexus LS sedan. The company claimed this hardware was designed with a low center of gravity as one of its priorities. The body shell was constructed from a combination of high-strength steel, aluminum, and carbon fiber reinforced plastic. The rear-drive LC ran on a 113-inch wheelbase, was 187.4 inches long, and stretched 75.6 inches wide. Curb weight was a substantial 4280 pounds.

The LC 500's extra-long hood hid a 5.0-liter naturally-aspirated 32-valve V8 engine. It was rated at 471 bhp and 398 pound-feet of torque. Horsepower peaked at a lofty 7100 rpm, with a redline set at 7300 rpm. The engine mated to a 10-speed automatic transmission and was fitted with a standard active exhaust system that allowed the driver to select the level of exhaust sound. Its rambunctious exhaust added a surprising muscle car vibe. Lexus claimed the LC 500 would accelerate from 0 to 60 mph in 4.4 seconds.

The standard wheels were 20-inch cast-alloy units, though optional forged wheels in 20- and 21-inch sizes were available too. All tires were run-flats. Chassis highlights included multilink independent suspension front and rear and beefy steel-disc brakes. An active rear steering system was part of the optional performance package. As a 2+2 coupe, the LC 500 was dramatically styled and luxuriously appointed inside. However, that style came at the price of nearly nonexistent rear-seat legroom.

At introduction, the LC 500 was priced from $92,000. Individual extras included a cold weather pack with a heated steering wheel and windshield deicer, color heads-up display, and a limited-slip rear differential. Option groups were a Touring Package, Sport Package, Sport Package with Carbon Fiber Roof, Performance Package, and Mark Levinson-brand audio.

In a world where the idea of automotive luxury was shifting to SUVs, grand touring coupes still excelled at delivering drama. The Lexus LC 500 could easily cost north of $100,000, but it packed exotic looks, beautifully finished interior, and, of course, ample V8 power.

PORSCHE

911 GT2 RS 2018

The GT2 RS was the most powerful and expensive member of Porsche's 911 lineup for 2018, and it was also one of the most ferocious, uncompromising 911s ever built. Racetrack prowess was the GT2 RS's overriding focus, so Porsche prioritized trimming weight wherever possible. Carbon fiber was used extensively; sound insulation was deleted; and a $31,000 Weissach Package added weight-saving gear such as magnesium wheels, carbon anti-roll bars and links, a carbon fiber roof panel, and even lighter-weight carpet. The air-conditioning and infotainment systems could also be deleted if the buyer desired.

The GT2 RS's powerplant was a real monster: a twin-turbocharged 3.8-liter six that made 700 horsepower and was paired solely with a 7-speed automated manual transmission—unlike previous GT2s, a traditional manual transmission wasn't offered. Other high-powered 911s used all-wheel drive to tame power delivery, but the GT2 RS was rear drive only. Performance figures were jaw-dropping: 0 to 60 mph in 2.7 seconds, 0 to 124 mph in 8.3 seconds, and a top speed of 211 mph. Plus, the finely tuned chassis and suspension were up to the task of handling the formidable horsepower.

GT2RS

Large carbon-ceramic brake discs, cooled by ducts in the front trunklid and fender extractor vents, provided outstanding stopping power. Active rear-axle steering provided better responsiveness at low speeds and improved stability at high speeds. The GT2 RS proved its potential at Germany's famous Nürburgring-Nordschleife by lapping the 12.8-mile course in 6:40.3 seconds—a record for a road-legal vehicle on that track. Naturally, these superhero capabilities didn't come cheap—the base price of the GT2 RS was $293,200.

MCLAREN

SENNA 2019

The McLaren Senna was a testament to the company's racing heritage. Founder Bruce McLaren was a New Zealand born driver who started building English racecars in 1963. Legendary Brazilian driver Ayrton Senna raced for McLaren from 1988 to 1993, racking up three Formula One world championships. He was tragically killed while driving for another team in 1994.

The McLaren Senna was built as an uncompromising track car that could set fast lap times, yet was street legal. The Senna was based on the 720S. While the 720S was comfortable and docile for a mid-engine supercar, the track-focused Senna had firmer suspension and lacked sound insulation. All unnecessary weight was eliminated. The 2641- pound car was powered by a 789 horsepower twin-turbocharged 4.0- liter V8 and was capable of a top speed of 208 mph. Acceleration times were: 0-62 mph (100km/h) in 2.8 seconds, 0-124 mph (200km/h) in 6.8 seconds, and 0-186 mph (300km/h) in 17.5 seconds. Strong brakes are important to quick lap times and the Senna could come to a complete stop from 62 mph in less than 100 feet. Each carbon-ceramic brake disc took seven months to create.

The Senna's aggressive styling was an exercise in "form follows function" with an emphasis on generating downforce to keep the car stable at high speeds. McLaren claimed the Senna generated 1763 pounds of downforce at 155 mph.

The cost of the Senna was around $1 million, but the 500-car quota was sold out before production began.

FORD

MUSTANG SHELBY GT500 2020

In the sixties, former race-car driver Carroll Shelby modified Ford Mustangs and sold the result under his own name. The Shelby name graced special production Mustangs over the years, but recently had been on hiatus. That changed with the introduction of the 2020 Ford Mustang Shelby GT500 powered by a 760 hp supercharged 5.2-liter V8. Backing that was a seven-speed dual-clutch automated-manual transmission, an adjustable suspension, and Brembo brakes, all combining to make for a Mustang that is a ferocious performer on both the drag strip and the racecourse. Ford claimed 3.3-second 0–60 mph times and quarter-mile times of less than 11 seconds.

TOYOTA SUPRA 2020

Ending a 22-year hiatus, the Toyota Supra returned to showrooms worldwide in 2020. The 2-seat sports car featured a turbocharged 6-cylinder engine mated to an 8-speed automatic transmission. A manual transmission was not offered. Codeveloped with German luxury maker BMW, the Supra was assembled alongside BMW's Z4 coupe and convertible in Graz, Austria.

In 2020, the Supra's engine produced 335 horsepower. A power bump would later take the engine to 382 horsepower. For 2021, a 4-cylinder entry-level Supra was on offer.

TOYOTA
Supra
GR TOYOTA Supra

GR TOYOTA Supra

POLESTAR

1 2020

A spin-off of Swedish luxury carmaker Volvo, Polestar's mission was to build and sell only high-end electrified performance vehicles. Polestar 1 was the new brand's debut offering. A compact coupe, the "1" was a plug-in hybrid combining a turbocharged and supercharged 4-cylinder engine with a pair of electric motors for total system output of 600 horsepower.

Built at Polestar's factory in Chengdu, China, the Polestar 1 was capable of reaching 60 mph in under four seconds. And, when fully charged—and conservatively driven—the coupe could travel more than 50 miles without using any gasoline.

Polestar 1

RIMAC

NEVERA 2022

With an available 1914 horsepower, the all-electric Rimac Nevera screeched to 60 mph from a stop in under two seconds. The product of a Croatian "sports car" maker with ties to Bugatti, the Nevera was first seen at the 2019 Geneva Auto Show. Power for the Nevera's four electric motors was supplied by a 120-kWh lithium-ion battery which provided a claimed 340 miles of driving range. Production of the $2.4 million sports car was limited to 150 vehicles.